The Instrumental Spectrometric and Spectroscopic

Analysis of Natural Food Flavourings

Volume I

A Primer

Dr. Nick Winstone-Cooper

First published by Ellixia Publishing Limited in 2020
www.ellixia.com

978-1-71698-733-5

Acknowledgements: Most images and diagrams have been created by the writer and those from other sources are acknowledged with thanks for the permission to use them. All photographs of the scientists are out of copyright but where the source is known it is acknowledged.

Written for

and the Welsh Government's educational Hwb.

Dr. Nick Winstone-Cooper studied chemistry and physics at Cardiff University before completing postgraduate research in nuclear chemistry, focusing on the creation of radioactive complexes of macrocyclic phosphines for application as heart and bone imaging agents in cancer diagnosis.

He worked extensively in the United Kingdom, North America, Italy and the Republic of Korea before moving into education and education consultancy.

Series introduction

This series of three volumes is an introduction to the structural analysis of organic compounds with an emphasis on the determination of the molecular structure of natural food flavourings.

The three volumes provide all the knowledge, understanding and skills to interpret elemental and spectral data at both A – Level Chemistry and first year university levels.

The three volumes comprise:-

I: A primer on atomic structure and the theory and principles behind the analytical techniques ;

II: The application of the theory to determine and confirm the structures of small molecules;

III: The third volume uses the understanding of the principles and application of these techniques to investigate and determine the structure of natural food flavourings.

Volume I

This volume describes and discusses those discoveries, principles and techniques necessary to understand the analysis of small molecules (discussed in Volume II) and natural food flavourings (addressed in Volume III).

It is unlikely that readers will come to this series without some prior knowledge but it is intended that this volume should also be accessible to all new to this field and it is written to be as close to a conversation between writer and reader as a one to one tutorial can be but in writing and with no prior knowledge expected of the reader beyond a general knowledge of the existence of elements and the conception that the atom is the smallest part of an element which still behaves as that element.

We start from the 19th century concept of the atom as a small, hard and indivisible particle about which we can learn nothing and continue through to the current view of the atom as comprised of a dense nucleus containing protons and neutrons with the electrons in identifiable regions outside the nucleus. We continue through the major discoveries about the atom to our ability to identify the precise arrangements of the atoms in the molecule and the interactions between the atoms.

It is intended to be accessible to all interested readers, assumes no more than secondary school mathematical knowledge and includes a glossary of commonly used terms which have specific meaning within physics and chemistry.

Table of Contents

Chapter V – Bonding Concepts and principles

Chapter VI – Infrared Spectroscopy and Mass Spectrometry

Chapter VII – Nuclear Magnetic Resonance Spectroscopy – Theory, Principles and Applications

Chapter VIII – Pen portraits

Introduction

Fifty Years When Physics and Chemistry Shook The world

The aim of this series is to guide the reader from no assumed knowledge of the structure of the atom through to an ability to determine the structure of natural food flavourings using a range of measurements and interpretations of this information.

Modern techniques of analysis include infrared spectroscopy, mass spectrometry and multi-nuclear magnetic resonance which allow us to determine and explain the precise arrangements of all atoms within a discrete molecule.

Looking at spectroscopic and spectrometric spectra can initially be extremely daunting but if a routine approach is developed and then adopted the spectra become easier to interpret and, if certain clues are sought, then analysis becomes much less challenging.

There are a number of steps to use:

- elemental composition calculations,
- consideration of atomic structure,
- infrared spectroscopic analysis,
- mass spectrometric analysis and
- analysis of multi-nuclear magnetic resonance spectra.

and this primer addresses those principles and applications in that order.

Progressing through the pages, we initially examine how to perform fundamental calculations since knowledge of the presence of specific elements, their relative proportions and the relative molecular mass of the molecules are essential before attempting to interpret spectrometric and spectroscopic data. We continue by looking at the series of discoveries that have led our view of the atom to change from the concept of a solid, indivisible ball to one where we know of the existence of the nucleus, the proton and neutron, the electronic distribution around the nucleus and the rather mind boggling conclusion that an atom, and so much of our body, is mainly comprised of empty space.

The following discoveries are discussed in their relevant order i.e.

- Goldstein's discovery of, positively charged, *canal rays*, later determined to be protons;
- J.J. Thomson's discovery of, negatively charged, *cathode rays*, later determined to be electrons, and his '*plum pudding*' model of the atom;
- Rutherford's discovery of the existence of the, positively charged, atomic nucleus;
- Nagaoka's concept of the electrons orbiting the nucleus as the planetary rings orbit Saturn;
- Aston's invention of the *mass spectrograph* and hence the discovery of *isotopes*, whose existence were yet to be explained, but whose existence were explained some years later by James Chadwick's discovery of the existence of the neutron;
- Moseley's determination of the significance of the number of nuclear protons which is now known as the *atomic number*;

- Niels Bohr's proposal for electrons orbiting the nucleus which depended on two other discoveries: the well known atomic emission spectra and Einstein's explanation of the *photoelectric effect* which depended on Planck's concept of electromagnetic radiation as small packets of energy known as *photons*, in other words that waves can also behave as particles.

- Louis de Broglie's suggestion that if waves can behave as particles then particles must able to behave as waves, leading to the concept of *wave – particle duality* which led on to:-

- Erwin Schrödinger's wave equation which, in turn led to the concept of electronic orbitals which built on Bohr's model but more explicitly and precisely predicted the probability of the location of electrons within each of Bohr's energy levels.

- The concept of the electronic distribution of the atom where they are located in Niels Bohr's orbits (energy levels) led on to G.N. Lewis' determination of the significance of the Noble Gas configurations.

- We also make much use of the concepts of ionic and covalent bonding as described in Linus Pauling's magisterial but very readable volume '*The Nature of the Chemical Bond*'.

- Later instrumental developments enabled the development of mass spectrometry, originally used in the discovery of isotopes, to become a routine tool in organic compound analysis.

- The most important technique though is nuclear magnetic resonance which enables an analyst to identify the precise location of all hydrogen and carbon atoms within a molecule. This technique depends utterly on the conception of the nucleus, the arrangement of electrons within the atom and the fundamental properties of a spinning charge.

In this manner we move from the initial conception of a hard, solid and indivisible atom to intimate knowledge of structure of the atom and its components (protons, neutrons and electrons) through to an ability to determine the structure of organic molecules of great importance due to the theory of quantum mechanics.

It is worth remembering that whilst some people believe that science is simply a matter of an inevitable series of sequential discoveries and that all inventions are waiting to be discovered but this is far from the truth. For example, the smart phone was not waiting around waiting to be invented and neither was the computer on which this series was written. Inventions are a combination of a number of serious but apparently and originally unrelated discoveries.

Science is better portrayed as trying to get through a rather large and complicated maze. There are many blind alleys and there can be many years between discoveries. Many of these discoveries are found by accident and at the time are viewed as mere curiosities which may or may not have a use and which cannot, at the time, be explained. This was the case, for example, with the photoelectric effect and notably with emission spectra. Many discoveries were entirely predicted but found by accident whilst some were made from specific hypotheses. For example, the wavelike nature of the electron, predicted by Louis de Broglie on the basis of Einstein's determination that waves could, under certain circumstances behave as particles, was confirmed entirely accidentally by Davisson and Germer and, separately and independently, by G.P. Thomson. Whilst Thomson was interested in the properties of the electron and deliberately examining them, Davisson and Germer were using electrons to investigate the crystalline structure of metallic nickel and produced the diffraction pattern of electrons, hence demonstrating their wavelike nature, as a side effect of their investigation. For this accidental discovery, Davisson received the 1937 Nobel Prize in Physics.

Likewise, the photoelectric effect was discovered by Heinrich Hertz who was experimenting with spark gap devices (which comprise two, closely spaced but separated electrodes designed to allow an electric spark to pass between them) and discovered that their reliability could be improved if ultraviolet light was shone on one of the two plates. Hertz was not investigating atomic structure as he was interested in radio. The consequences of his discovery combined with Max Planck's mathematical trick of considering energy as tiny, particulate, bundles of energy and not waves led Einstein to his determination that electromagnetic radiation can be both wavelike and particulate in nature. Although it has, so far, been impossible to reconcile his theory of relativity to quantum mechanics, the consequences of Einstein's work on the photoelectric effect to the formation of modern quantum theory cannot be overestimated.

To truly begin to understand these concepts it appears wise to suspend built in preconceptions about energy & matter and waves & particles as we are all taught them in schools and experience in ordinary life especially the artificial distinction between

Waves & Particles and Energy & Matter

To achieve a distinction between waves and particles is extremely difficult since we all know, from just visiting a beach, what we understand to be a wave and a particle such as a pebble on the beach. Our experience of both makes them appear to be completely distinct but that is merely our perception which is entirely usual.

Indeed this was the view held by all physicists until Einstein's explanation of the photoelectric effect which, discussed later, employed the concept that the impinging light on a metal could be viewed as a stream of discrete particles.

Similarly, energy was defined as the potential for an object to perform work (potential energy) or the energy of a moving object (kinetic energy) whilst mass was defined both as a property of a physical body and a measure of its resistance to acceleration when a net force is applied. They were viewed as entirely unconnected until Einstein, again, related them in his theory of special relativity when he derived his famous $E=mc^2$ equation.

It is, perhaps, better to remove all preconceptions and to start from scratch. If the reader finds this difficult then rest assured that all physicists also suffered from the same confusion and, indeed, Richard Feynmann (1965 recipient of the Nobel Prize in Physics) wrote *"If you think you understand quantum mechanics then you do not understand quantum mechanics"*. This is reassuring for all who find the concepts difficult which mean all of us and this is one reason that some concepts are reiterated in this book rather than just being introduced once and then left, with understanding assumed. Fortunately, chemists make use of the conclusions which provide great insights into the structure of the atom and molecules and use the conclusions for our own purposes and we escape the mathematics required to derive the conclusions of quantum theory.

Science is often portrayed as a matter of analysing data by cold people who work entirely objectively.

That this is nonsense is demonstrated by the recent furious debates about climate change etc; and scientists have all the emotions demonstrated by other people and can choose the data to support their own point of view.

It must also be borne in mind that even at the time of the discovery of what we now call protons and neutrons, some scientists still disputed the very existence of the atom and there were many angry arguments about whether or not atoms existed which seems quite extraordinary to us today. A similar matter occurred with *The Principle of the Conservation of Energy* which is taught to schoolchildren as a given fact. There were vicious debates for over fifty years between its proposal and its final acceptance.

Continuing, we encounter men who were orphaned at a young age (Fraunhofer) and yet managed to become educated and provide the world with incredible insights through to members of the aristocracy, Louis de Broglie. Everybody also has an image of the '*mad scientist*' exemplified allegedly by Albert Einstein. In actuality he was, for most of his life, very well dressed and well turned out and it was only towards the end of his life that he developed the image of '*the hair*'.

For all these reasons, I strongly believe that it is important to have some knowledge of the scientists whose discoveries and inventions we have discussed in this volume and there are pen portraits of the scientists involved in the discoveries relevant to our investigation here.

Scientists are not just men who spend their time poring over their test tubes and their calculations but it is interesting that all the pen portraits though are of men since it demonstrates immediately how it was dominated by wealthy and well connected men.

That this is the case is demonstrated by the knowledge that female students at Cambridge were not permitted to graduate until 1948 and even then were not permitted to attended a graduation ceremony, instead receiving their degree certificate by post.

As with all other disciplines that sexist attitude has not been just to the womens' loss but also to all of us since who knows what genius and insights we have lost due to the sheer unfairness of the university system of the time and the discrimination?

However, returning to the scientists and inventors who are relevant to this series, we have serious sportsmen, musicians, mountaineers, men who built their own cars and raced them and men who, prior to 1914, had collaborated in research, fought in the First World War on opposite sides and then returned to cooperating in research.

Henry Moseley was killed at Gallipoli whilst Hans Geiger served as a German artillery officer. Lester Germer became a renowned fighter pilot and Sir George Paget Thomson served in both the British Army and the Royal Flying Corps and yet, after the First World War Armistice, many went back to working together despite having been fought on opposing sides.

Many physicists were deeply religious as demonstrated by the recipient of the 1932 Nobel Prize, Werner Heisenberg, and known as the '*father of quantum mechanics*', who wrote that "*The first gulp from the glass of natural sciences will turn you in to an atheist but at the bottom of the glass God is waiting for you..*

In the Second World War, we also have Jews (Niels Bohr and Werner Heisenberg) and those opposed to the Nazis who only just managed to escape Nazi – occupied countries (Erwin Schrödinger and Max Born).

Many were gregarious and party – loving including a scientist (Wolfgang Pauli) who married a cabaret dancer as well as, again, Erwin Schrödinger who lived in a ménage à trois and whose wife and mistress accompanied him across the world. Schrödinger was a virulent anti – Nazi and was only able to escape with his life because friendly colleagues gave him a few hours warning that he was about to be taken by the Gestapo.

In the Second World War, a number of these scientists worked on the development of radar, including Ernest Marsden and Isidor Isaac Rabi, who following the conclusion of the Second World War, was unable to obtain a research position because he was Jewish and spent time working as a bartender before returning to postgraduate study. James Chadwick, the discoverer of the neutron, and other physicists worked on the Manhattan Project developing the allies' nuclear bomb and used their knowledge and insight to move the technology from insight into the structure of the atom to creating a weapon of mass destruction. Many of these scientists, especially Chadwick, returned after the war, exhausted and troubled that their curiosity and insights had taken them to the development of weapons which could destroy the entire world.

Science is also a truly multi-national discipline. The scientists whose lives are, very briefly, recorded in the pen portraits below include those who were from England, Scotland, Switzerland, Poland, Russia, The Netherlands, Denmark, Germany, Austria, France, Japan, New Zealand and the United States of America.

Whilst many came to work in Britain, Marsden went to New Zealand, Sir William Henry Bragg went to Adelaide and Lord Rutherford was born near Nelson on the South Island of New Zealand. Many of the scientists discussed in this volume went back and forth between Britain and America, Europe and Australia.

It is also worth mentioning the Nobel Prize Winners and the universities at which these discoveries were made. Of the Nobel Prize Winners relevant to this series, we have two sets of father and son laureates.

Sir William Henry Bragg and Sir William Lawrence Bragg were joint recipients of the 1915 Nobel Prize in Physics for their joint work on x-ray diffraction. The Braggs had caused a huge controversy for their discovery that sodium chloride (common or 'table' salt) was not a molecule (NaCl) but was actually a crystalline structure with alternating sodium and chloride ions but with the same empirical formula. This completely contradicted the notion that all compounds exist as molecules and the response by some scientists was vicious, once again contradicting the notion of cold and objective scientists. The *Bragg Controversy* raged for some years but their conclusions are now taught to schoolchildren as an example of ionic bonding and which demonstrates the properties of ionic compounds: solubility in water, strength and high melting points

Father and son, Sir Joseph John *J.J.* Thomson and Sir George Paget *G.P.* Thomson both received the Nobel Prize in Physics for their work on the electron. *J.J.* determined that the cathode rays were negatively charged particles (electrons) whilst *G.P.* determined that they also had wavelike properties

Meanwhile we also encounter Linus Pauling, who formalised the concepts of covalent bonding and derived the theory of electronegativity and who is also unique in that he was the recipient of two Nobel prizes in two completely different disciplines, the 1954 Nobel Prize in Chemistry and eight years later, for his work opposing nuclear weapons, the 1962 Nobel Peace Prize. Pauling was also one of the creators of the Pugwash Movement which works towards nuclear disarmament which, as itself as an organisation, was awarded the 1995 Nobel Peace Prize.

Scientific discoveries are never confined to one location. For example, although the most famous scientific academic institute is, perhaps, the Cavendish Laboratory in Cambridge, much of the British work in this subject was conducted at the universities of Liverpool, Manchester and Aberdeen demonstrating the strength of the British academic system. Indeed, given the number of discoveries there, Manchester is perhaps more important than Cambridge.

The following pages present, mainly one page, pen portraits of all the scientists who are referred to in this volume. The term *scientist* was coined in 1837 at the annual meeting of the British Association for the Advancement of Science. Until then, *science* was known as *natural philosophy* and it is, again, noteworthy that are all those portrayed are white men and there are several reasons for this.

Many British men involved in the study of chemistry and physics in 18th and early 19th centuries were of independent wealth and, rather than having to work for a living, were able to devote their time to their scientific interests.

Women of wealthy families were discouraged from acquiring formal education and, instead, were taught at home whilst their brothers, even if younger, were educated at Eton, Harrow or Westminster and then at Oxford or Cambridge. Teenage girls and young, unmarried women, were expected to develop skills in painting, art, music and conversation whilst moving in the circles of polite society before marrying and continuing the family lines of their father and their new husband.

Even in 20th century America and Britain, in only certain circumstances could black men attend universities and then, in America, could only study at specific colleges.

These colleges did not have the funds to invest heavily in science and technology and many struggled to survive from academic year to academic year. That this discrimination persisted until relatively recently is also demonstrated by the, only recently publicised, role that African – American women of huge mathematical ability played in the, 1969, landing on the moon but who had never been allowed to study at college.

That the scientists whose discoveries are discussed in this series were all white, many were very religious and few grew up in deprived circumstances but suffered discrimination provides revealing insights into the society in which they lived and worked and from an unlikely perspective.

The scientists portrayed in the Chapter VIII of this volume are only those who played a highly significant role in the overall story of the discovery of the structure of the atom and the instrumental methods relevant to this series of books. They were not the only ones.

This primer is intended to provide everything needed to apply knowledge and skills to, initially confusing and frankly quite off-putting, chemical and physical measurements and it is intended to be an enjoyable read.

I hope that those objectives have been achieved.

Chapter I

Fundamental Elemental and Molecular Concepts

This chapter considers:-

- The determination of the elemental composition of molecular compounds;
- The calculations of empirical and molecular formulas of molecular compounds;.
- Aspects of the electromagnetic spectrum relevant to molecular structure determination; and
- Definitions pertinent to this series such as *work, energy, charge* and *magnetism* and a reminder of the nature of the *electromagnetic spectrum*.

These concepts are fundamental to the following pages and it is imperative that they are grasped in order to fully understand the spectrometric and spectroscopic techniques discussed later and to learn to apply.

This is why we now spend some time working through them.

Elemental analysis

When conducting spectral analysis, the only actual *facts* available to chemists are the relative atomic masses of the elements (A_r), elemental composition, the relative molecular mass (M_r) and the exact measurements from the spectral data. All other analyses are interpretations, calculations and inferences from comparisons with other molecules. The most fundamental data is the elemental composition and the relative molecular mass (M_r). No spectral interpretations can be consisted to have any validity without that basic and essential information.

The elemental analysis of a compound may appear to be boring and trivial but, in reality, it is an extremely skilled task given that the sample analysed is rarely of a mass greater than 0.5mg and is such a small quantity it can hardly been seen. Measuring such a small mass precisely is extremely difficult and consequently the elemental composition is measured a number of times and mean results calculated. This means that, occasionally, this means that the total elemental composition does not always total 100% and 99.95 – 100.05% is acceptable. Traditionally, organic compounds were analysed for their carbon, hydrogen and oxygen content via combustion analysis. The carbon was collected as carbon dioxide, hydrogen as water and oxygen was calculated by difference. Nitrogen was determined separately by the Kjeldahl technique whilst sulfur was oxidised to sulfur dioxide which, dissolved in water, could then be titrated, having formed sulfuric acid. Halogens were determined by methods involving fusion with a known mass of sodium metal and gravimetric (weighing) analysis of the product.

The original combustion analytical method was established by the renowned French chemist, Joseph Gay-Lussac (pictured below left), who used his method to establish the elemental composition and hence molecular

formula of water. Until then, although water was known to contain both hydrogen and oxygen, it was believed that the molecular formula of water was believed to be HO, not H_2O. Gay-Lussac's method was further developed by Justus von Liebig (pictured right) who was responsible for the design of much laboratory glassware, notably the essential and indispensable Liebig condenser used in every organic chemistry laboratory to this day. Liebig combined

Gay-Lussac's methods to form an *'analytic train'* of measurements enabling the determination of the C, H, O content of a molecule in one experiment. More modern instrumental methods still involve combustion of almost invisible quantities of a sample with detection by infrared methods but the essential principles remain unchanged.

It is a mistake to dismiss these methods as relatively unsophisticated when compared to modern instrumental techniques such as infrared and multinuclear nmr spectroscopy and mass spectrometry. Without precise, reliable and accurate elemental composition data all other measurements and interpretations have no validity whatsoever. It is also worth understanding the difference between spectroscopic and spectrometric measurements. Spectroscopy involves the measurement of the effects of electromagnetic waves e.g. light whilst metric is simply a measurement and *spectro* refers to a range and *metric* means a range. There are many excellent textbooks explaining the theory of all techniques and so the following pages simply describe the principles relevant to the analyses of the compounds investigated in this series.

Empirical and Molecular Formula Calculations

Although difficult to master initially, calculating the empirical formula and the molecular formula can become routine if a set procedure is followed. This will be demonstrated by calculating the empirical formula and the molecular formula of butanoic acid and then three other compounds (propanol, propanoic acid and cyclohexene) which are discussed later in the context of spectrometry and spectroscopy. It is worth giving a number of examples as the data is entirely fundamental to all other measurements.

As a reminder, the *empirical formula* is the number of atoms of each element present in their lowest possible ratio whilst the *molecular formula* states the exact number of atoms of each element present in the molecule. Since many compounds e.g. inorganic compounds are not molecules it is becoming more common to use the term *formula mass* but for organic compounds which we are exclusively dealing with the conventional term, relative molecular mass (M_r), is retained.

Calculating the empirical formula

Step 1: The elemental composition of butanoic acid is C: 54.50% H: 9.17% O: 36.33%

Step 2: Divide each percentage composition by the relative atomic mass of the element:

	C	H	O
	$\dfrac{54.50}{12.00}$	$\dfrac{9.17}{1.01}$	$\dfrac{36.33}{16.00}$
=	4.54	9.08	2.27

Step 3: Divide the results by the lowest number which in this case is 2.27 (oxygen):

	C	H	O
	$\dfrac{4.54}{2.27}$	$\dfrac{9.08}{2.27}$	$\dfrac{2.27}{2.27}$
=	2.00	4.00	1.00

This means that the empirical formula is C_2H_4O.

Calculating the molecular formula

Continuing the numbering from the previous steps above

Step 4: The relative molecular mass (Mr) of butanoic acid is 88.08 g mol^{-1}.

Step 5: Calculate the relative molecular mass from the empirical formula.

The Mr of a fragment of empirical formula, C_2H_4O, is 44.04 g mol^{-1}.

The Mr of the actual molecule is exactly twice that value so the molecular formula is $C_4H_8O_2$.

Since the compound is a weak acid, this means that the structural formula must be:

Three further examples follow in order to build confidence and refer to molecules analysed later.

Propanol

Step 1: The elemental composition of propanol is C: 59.92%, H: 13.45%, O: 26.63% and the molecule has a relative molecular mass of 60.08 g mol^{-1}.

Step 2: Divide each percentage composition by the relative atomic mass of the element:

	C	H	O
	$\dfrac{59.92}{12.00}$	$\dfrac{13.45}{1.01}$	$\dfrac{26.63}{16.00}$
=	4.99	13.32	1.66

Step 3: Divide the results by the lowest number which in this case is 1.66 (oxygen):

	C	H	O
	$\dfrac{4.99}{1.66}$	$\dfrac{13.32}{1.66}$	$\dfrac{1.66}{1.66}$
=	3.01	8.02	1.00

This means that the empirical formula is C_3H_8O.

Calculating the molecular formula

Continuing the numbering from the previous steps above

Step 4: Note that the relative molecular mass (Mr) of propanol is 60.08 g mol^{-1}.

Step 5: Calculate the relative molecular mass from the empirical formula.

The M_r of a fragment of empirical formula, C_3H_8O, is 60.08 g mol^{-1}. This means that the molecular formula is the same as the empirical formula and so is C_3H_8O.

There are two possible structures for the molecule since the oxygen atom can sit on one of the terminal carbons or on the middle carbon atom.

propan-1-ol   **propan-2-ol**

Note that there is no such molecule as propan-3-ol as simple flipping over reveals it to be the same as propan-1-ol.

The instrumental, spectrometric and spectroscopic, methods can be used to distinguish between these two possible structures and since both molecules are small and very common they are extremely useful in developing and understanding the principles behind the analysis.

These isomers are used in deeper discussion later in this volume.

Propanoic acid

Step 1: The elemental composition of propanoic acid is C: 48.61%, H: 8.18%, O: 43.21% and the molecule has a relative molecular mass of 74.06 g mol^{-1}.

Step 2: Divide each percentage composition by the relative atomic mass of the element:

	C	H	O
	$\dfrac{48.61}{12.00}$	$\dfrac{8.18}{1.01}$	$\dfrac{43.21}{16.00}$
=	4.05	8.10	2.70

Step 3: Divide the results by the lowest number which in this case is 2.70 (oxygen):

	C	H	O
	$\dfrac{4.05}{2.70}$	$\dfrac{8.10}{2.70}$	$\dfrac{2.70}{2.70}$
=	1.50	3.00	1.00

Atoms only exist as whole numbers then multiplying by two means that the empirical formula is $C_3H_6O_2$.

Calculating the molecular formula

Continuing the numbering from the previous steps above

Step 4: Note that the relative molecular mass (Mr) of propanoic acid is 74.06 g mol^{-1}.

Step 5: Calculate the relative molecular mass from the empirical formula.

The Mr of a fragment of empirical formula, $C_3H_6O_2$, is 74.06 g mol^{-1}. This means that the molecular formula is the same as the empirical formula and so is $C_3H_6O_2$.

The only plausible structure of the molecule is

and this is confirmed by the instrumental, spectrometric and spectroscopic, methods discussed later in this volume.

It also reinforces the chemical data such as that the crystalline solid dissolves in water and creates an acidic solution and reacts with, neutralising, alkaline solutions.

Cyclohexene

Step 1: The elemental composition of cyclohexene acid is C: 87.70%, H: 12.30% and the molecule has a relative molecular mass of 82.10 g mol^{-1}.

Step 2: Divide each percentage composition by the relative atomic mass of the element:

C	H
$\dfrac{87.70}{12.00}$	$\dfrac{12.30}{1.01}$

$=$ 7.31   12.18

Step 3: Divide the results by the lowest number which in this case is 7.31 (carbon):

C	H
$\dfrac{7.31}{7.31}$	$\dfrac{12.18}{7.31}$

$=$ 1.00   1.66

Atoms can only exist as whole numbers so this means that we have to multiply the proportions:

Multiplying by   **gives us**

Multiplying by	C	H
2	2.00	3.32
3	3.00	4.98

Rounding up gives us the empirical formula C_3H_5.

Calculating the molecular formula

Step 4: Note that the relative molecular mass (Mr) of propanoic acid is 82.10 g mol^{-1}.

Step 5: Calculate the relative molecular mass from the empirical formula.

The Mr of a fragment of empirical formula, C_3H_5, is 42.05 g mol^{-1}.

Since this is half of the relative molecular mass of the molecule, this leads to the obvious conclusion that whilst the empirical formula is C_3H_5, this means that the molecular formula is C_6H_{10}.

The only plausible structure of the molecule is

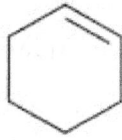

and this is confirmed by the instrumental, spectrometric and spectroscopic, methods discussed later in this volume. Note that it is irrelevant where the C = C bond is drawn as a simple rotation makes all possible structures chemically and magnetically equivalent.

Definitions

It is very important that scientists know exactly what each is referring to, for two reasons, clarity and brevity. It is, sometimes, said that scientists want people to be *blinded by the science*. This is not correct and it is not a deliberate attempt to exclude anybody. It is merely an attempt to refer to certain concepts which then require no further explanation.

There can be confusion about terms which have very precise meanings in science, again for clarity, but which have multiple meanings in every day usage. Four examples are **work**, **energy**, **charge** and **magnetism**.

Work

We all know that **Work** can mean, in everyday English language, paid work, classwork, homework etc; In physics, though, **Work** has a very precise definition:-

Work is performed when **something** is made to move by something else.

There is no reference to how the **something** is made to move or even what the **something** is and it can, therefore, cover movement due to impact of particles (*mechanics*) such as the collision of the white ball with other balls on a pool table and also action at a distance by *fields* (magnetic and/or electric) which is demonstrated by the repulsion or attraction between two bar magnets depending on the orientation of the magnets and hence their poles.

Energy

Given the above definition of **work** we can now define the term **energy** which does sound a bit vague but is actually very precise. Again, we all know the everyday meaning such as when tired but asked to do something we say *"Oh, I don't have the energy . . ."*

However, in physics:- **Energy is the ability to perform work.** In other words, possession of **energy** is the possession of the ability of *something* to move *something else*. Again the *something* and *something else* are undefined.

Energy is the basis of everything and, whatever pupils learn in school, there are only two forms of energy:

- There is energy associated with movement which is known as **kinetic energy**;
- All other forms of energy are stored and provide bodies with the *potential* to perform work and are traditionally known as **potential energy**. These types are now more commonly described as *stored energy* and include chemical energy, nuclear energy, the energy stored in a compressed spring or a taut rubber band (elastic energy), the energy we obtain from food (also chemical energy) etc; but the clearest example is *gravitational potential energy* (G.P.E.) which explains why a ball will drop down a staircase or a ball kicked over a cliff will fall to the ground.

We also have magnetic and electrical energy which brings us on to electromagnetic radiation which is a means of transferring energy from one location to another, remote, location and is highly significant for the rest of this series. Later pages discuss the movement of electrons between energy levels and require either the input of energy to allow transition from one level to a higher one, ultimately leading to ionisation, or releasing energy when transitioning from a higher to a lower energy level.

Charge

The concept of charge has been described since at least the time of the ancient Greeks who knew the effect of rubbing amber on a material and then attracting or repelling other materials. This is replicated in every schoolchild's experience of rubbing a plastic rule on a sleeve and picking up small pieces of paper, repelling another charged ruler, made of different material, and also diverting a stream of water from a tap.

Scientists use the term **charge** so often that it is easy to forget exactly what the term means and why there are two forms of charge.

Put simply, and as the ancient Greeks described it,

Charge is the ability of a material to **_attract_** or **_repel_** another material

Since there are two actions of **_charge_**: _attraction_ and _repulsion_ they became referred to both as North & South and positive & negative. This is now understood to be due to _static_ electricity.

Charge has a special property however and that is its ability to perform work at a distance i.e. to make another object move. This _work_ is to attract or repel another charged substance but at a distance without any contact as in bar magnets whose N – N and S – S poles repel each other but N – S poles attract. This led to the concept of fields where the invisible _lines of force_ are drawn with an arrow as in the familiar pictures of bar magnets attracting or repelling each other depending on their arrangements. All school children are familiar with visualisation of the field using a bar magnet and iron filings which reveals the lines of force as shown below.

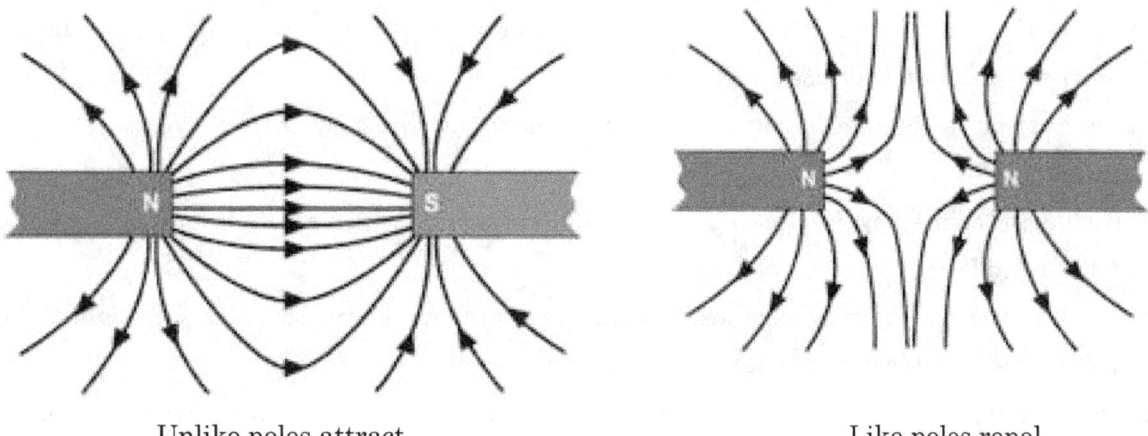

Unlike poles attract   Like poles repel

This also leads us to the definition of a _neutral_ material which is one which is unaffected by a charged object.

As an aside, when _current_ electricity was discovered and it was determined that it constituted a flow of these charges, circuit diagrams were drawn as if the flow of the current was from positive to negative and hence was described as a flow of positive charges.

We now know that the flow of charges is actually the movement of negatively charged electrons and the current flows in the opposite direction but, for consistency with the many decades of drawn circuit diagrams, circuit diagrams are still drawn in that way.

Magnetism

Magnetism is a physical phenomenon which results in attractive and repulsive forces between objects. All materials possess magnetic properties. There are three types of magnetism: *paramagnetism*, *diamagnetism* and *ferromagnetism* and the phenomenon arises from the electrons surrounding the nucleus which as a spinning charge create a magnetic field.

Diamagnetism occurs in all materials and is caused by the tendency of a material to oppose, and hence be repelled by, an applied magnetic field. Diamagnetism occurs in materials which contain no unpaired electrons. *Spin pairing* of electrons is discussed later in this volume.

Paramagnetism is exhibited by substances that resonate with, and strengthen, an applied external magnetic field. They are materials which contain an unpaired electron in each atom and create their own field which cannot be cancelled out by the non-existent fellow electron and creates a *magnetic moment*. This unpaired electron is free to align its magnetic moment in any direction and does exactly that in the absence of an applied external magnetic field. The magnetic field of the Earth is far too weak to have any effect however, in the presence of an applied external magnetic field, these magnetic moments will tend to align themselves in the same direction as the applied field and hence reinforce and strengthen it.

Ferromagnetism is exhibited by the common metals iron, cobalt and nickel as well as by a handful of the rare earth metals and some alloys. This phenomenon is caused by the presence of an unpaired electron but in these materials there is also a tendency for these electrons' magnetic moments to align themselves permanently to create a lower energy state.

The specific details of the movement of an electric charge is of extreme importance:

- We now know that charged materials possess a surplus or a deficit of electrons and this means that they are, respectively, negatively or positively charged;
- An electric current is the movement of a stream of electrons which by their very movement create a magnetic field.
- An electric field is created by the presence of two different metals which are of different electronegativity which is discussed later;
- A charged particle can be moving whether changing its location or merely rotating (*spinning*) and nuclear spin is also discussed later in this volume;
- All particles rotate, *spin*, at even the lowest temperatures so magnetic fields are always created;
- In most cases the magnetic fields cancel each other out;
- There is a deep connection between electric charges and magnetic fields since the magnetism is created by the movement of the charge;
- The combination of electric fields and magnetic fields lead to the formation of electromagnetic radiation e.g. light.

All of this is highly relevant to the discoveries we discuss and investigate later in this volume and so it is worthwhile refreshing our ideas about the electromagnetic spectrum which is highly pertinent to our investigations.

The Electromagnetic Spectrum

It is worthwhile, briefly reconsidering the electromagnetic spectrum and when particular wavelengths of light can impact on chemical compounds and inform us about their structures. Just as all countries put their nation in the centre of a world map, we place the visible portion of the electromagnetic spectrum in the centre which is not mathematically correct but all regions are important and of use, as demonstrated below.

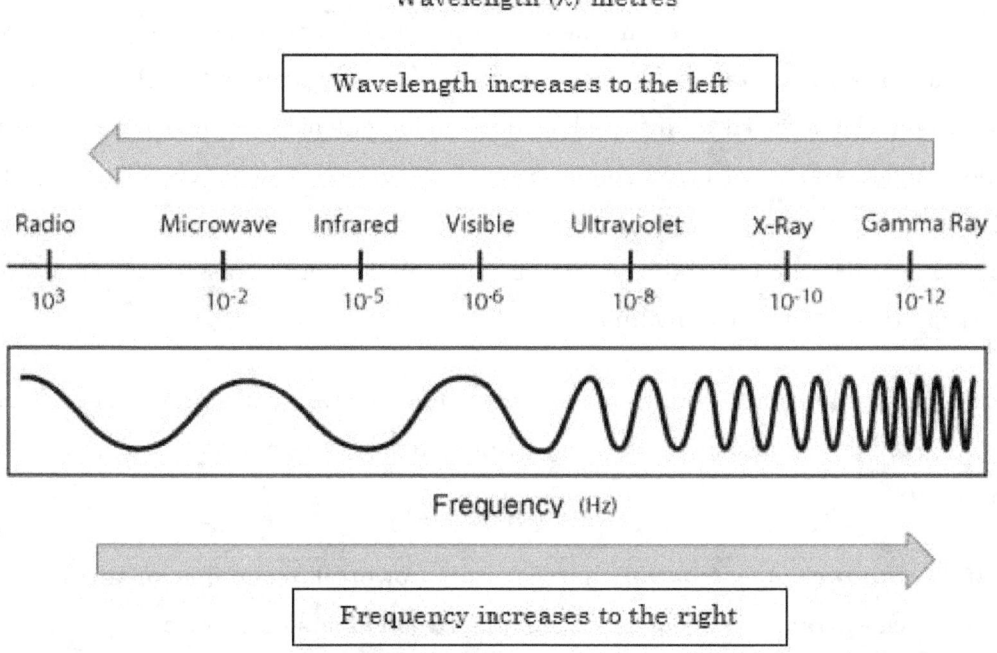

Applications of the electromagnetic spectrum and an approximation of the size of the wavelengths

Region	Scale	General Applications	Scientific Applications
Radio	Buildings	Radio and TV transmissions	Multinuclear NMR spectroscopy
Microwave	Humans	Radar, mobile signals & heating food	Rotational analysis of molecules and air pollution analysis
Infrared	Butterflies	Remote controls and thermal imaging	Determination of functional groups and elemental composition
Visible	A fine needle point	Human Vision	Colours and the discovery of optical activity
Ultra violet	Bacteria and viruses	Suntans, Vitamin D production	Electronic transitions within metals, determination of oxidation states and water pollution analysis
X-Ray	Atoms	Medical assessments	Determination of the impurities in metals
Gamma ray	Atomic nuclei	Radiotherapy	Nuclear chemistry, astrophysics and geochemical analysis

Later in this volume we will investigate describe the application of infrared radiation, radio waves and visible light as well as, in some cases, the use of electric and magnetic fields. In the following pages we will investigate three techniques, infra red spectroscopy; mass spectrometry and nuclear magnetic resonance spectroscopy.

Infrared spectroscopy is a technique for the determination of the presence or absence of functional groups.

A *functional group* is an atom or group of atoms within a molecule that behave similarly wherever they appear in different compounds. This means that even when other parts of the molecule are quite different, certain functional groups tend to react in certain ways. For our purposes these include $O - H$, $C = C$ and $C = O$ and $C - O$ bonds which can all be detected by infrared spectroscopic analysis, enabling the confirmation of the structures of alcohols, alkenes, aldehydes and ketones as well as carboxylic acids and esters. Quite often it is not capable of distinguishing between isomers (molecules of the same molecular formula but different structures e.g. the isomers propan-1-ol and propan-2-ol) and this is where mass spectrometry and nuclear magnetic resonance spectroscopy become invaluable.

Mass spectrometry involves the application of electric fields to accelerate ions and magnetic fields to deflect them i.e. make the charged particles slightly change direction in order to separate them and determine their relative mass/charge ratio.

Nuclear magnetic resonance spectroscopy is the most powerful technique of all as it allows the determination of the precise arrangements of hydrogen and carbon atoms in a molecule.

To make use of the information all these techniques provide we need to understand the structure of the atom.

The following chapter discusses the various experiments and calculations that led from the concept of the atom being a hard and indivisible particle about which nothing could, apparently be known, to the determination that all matter has an associated wave and all waves can be associated with matter and some high energy waves can, indeed, produce matter. The concept of the wave – particle duality of matter is discussed in the next chapter and is a hard concept to accept and understand by all of us including the physicists who discovered it.

Chapter II

Discoveries about the structure of the atom

It is extraordinary to consider it now but many renowned physicists of the 19th century denied the entire existence of the atom whilst others still presumed it to be solid and indivisible even whilst the structure of the atom was beginning to be discovered.

In order to understand the changing view of the atom, it is essential to fully understand how Edwardian physicists' concept of the atomic structure became that shown below:

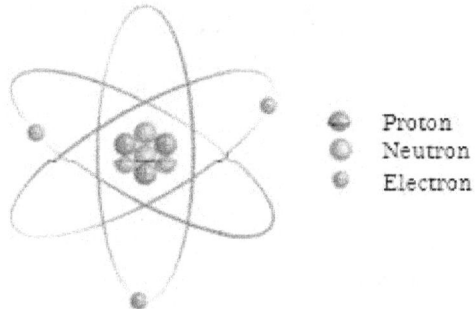

Proton
Neutron
Electron

Until the late 19th century, the atom was regarded as a solid, indivisible particle and many physicists regarded physics as a discipline to be almost complete. This was not out of arrogance but out of humility and, many being highly religious, meant that they believed that there were significant parts of nature which were beyond human understanding and analysis.

This changed rapidly, however, in 1886 when Eugen Goldstein discovered *anode rays*, when Henri Becquerel discovered *natural radioactivity* in 1896 and J.J. Thomson discovered *cathode rays* in 1897 and subsequently discovered that, although of the opposite charge to anode rays, they actually behave very differently.

This chapter considers the:-

- Discovery of *anode rays* and *cathode rays* leading to Thomson's '*Plum Pudding*' Model of the atom;
- The discovery of *isotopes* and the *nucleus* and the significance of the number of protons and the concept of the *atomic number* by Henry Moseley.

These discoveries take us to the view of the atom in Edwardian times but also include the discovery of the *neutron* (which occurred in 1932) before we then consider the Bohr model of the atom and the wave nature of particles in subsequent chapters.

The Discovery of Anode Rays and Cathode Rays

 Anode rays were first observed when a perforated cathode (negatively charged) and an anode (positively charged) were subject to a high electrical potential (several thousand volts) in a gas discharge tube as shown below. Very faint luminous rays were observed extending from the holes in the rear of the perforated cathode and travelling towards the anode. Eugen Goldstein (pictured left) termed these rays *Kanalstrahlen* (channel rays) as they were emitted from the holes or channels of the cathode. Since they were attracted to the negatively charged anode, these streams of particles were clearly positively charged and Goldstein had discovered the *proton* although this was not explicitly recognised at the time.

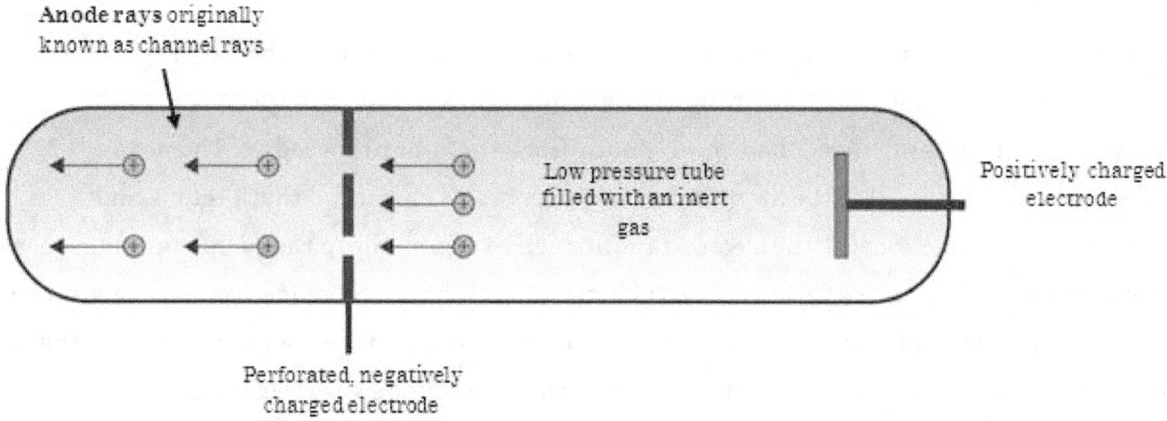

In 1897, Sir J.J. Thomson (right) performed a similar experiment (as shown below) and, placing a screen which would fluoresce (glow) when impinged by charged particles, observed cathode rays which, being attracted to the positively charged cathode, had to be negatively charged particles.

We now call them electrons.

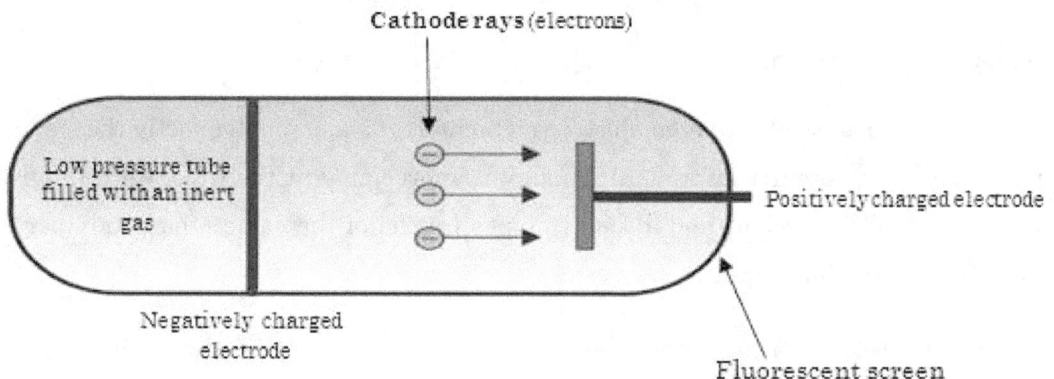

We can combine these two experiments as shown below.

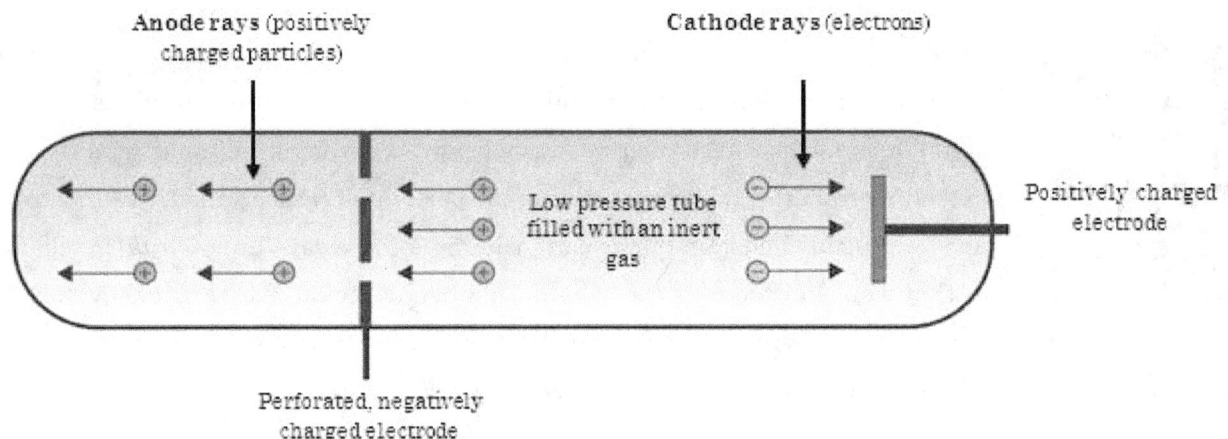

The discoveries by Goldstein and Thomson completely contradicted the notion of a hard, indivisible atom. Since both the, positively charged, protons and the, negatively charged, electrons (as we now term them) had to originate from the atom this led to Thomson's famous *plum pudding* model of the atom (right) in which it was envisaged that protons and 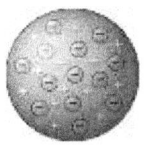 electrons held each other together by electrostatic attraction and made up the complete volume of a solid atom which had to be neutral since atoms were not affected by a magnetic field and passed straight through on its path as if the electric field was not present. This also naturally led to the logical assumption that the number of positive particles and the number of negative particles in an atom were the same.

Thomson went one step further though and placed his apparatus in a magnetic field. This part of the experiment was really quite simple to perform since the gas discharge tube was merely placed between the two poles of an external magnet. Thomson observed that the cathode rays were deflected (their direction of motion altered) when passing through a magnetic field and he was able to calculate the mass of the electron. In 1907, it was observed that the anode rays could be deflected by a magnetic field in a similar way.

Most significantly, however, whilst the cathode ray beam remained uniform and simply changed its route when traversing the magnetic field, the anode ray beam was not deflected uniformly but split into a number of rays which were deflected to different extents rather analogous to the splitting of white light by a prism into the colours of the rainbow. This implied that all cathode rays were identical but that the anode rays were not all exactly the same.

We can now explain Thomson's experimental results on the basis that:-

- Cathode rays are negatively charged electrons which are, of course, all exactly the same.
- The positively charged particles and, although not appreciated at the time, were the first indication of the existence of the nucleus and also of isotopes (nuclei of the same atomic number and so of the same positive charge but different masses).

This created a further puzzle since, by classical mechanical physics calculations, the masses of the particles constituting the anode rays could be calculated. It was consistently determined that the lightest particle was 1840 times greater than the mass of the electron. The **proton** had been discovered although the real significance of its role in atomic structure was not resolved until Rutherford's work in 1911.

The Discovery of the Existence of the Nucleus

In that year in Manchester, Rutherford assisted by, the soon to be famous in their own right, Hans Geiger and Ernest Marsden performed the classic '*gold foil*' experiment that led to the discovery of the existence of the atomic nucleus.

Hans Geiger   Lord Rutherford   Ernest Marsden

In this experiment alpha particles (helium nuclei $^4He^{2+}$), which had been discovered as a consequence of the discovery of radioactivity, were aimed at a piece of gold foil.

The entire apparatus was contained within a circular fluorescence screen which lit up briefly when hit by an alpha particle. It was predicted that if the atom was indeed completely solid then the alpha particles would be absorbed upon collision with an electron but repelled if they collided with a proton and none would reach a detector placed directly behind the gold foil.

This prediction is precisely what did *not* occur.

The result of the experiment was that the vast majority of the alpha particles passed straight through the gold foil. Most were deflected and passage through the gold foil changed their angle, but not the overall direction, of travel and reached the detector whilst a tiny fraction bounced off the foil and returned in various directions back towards the source of the alpha particles as illustrated below:

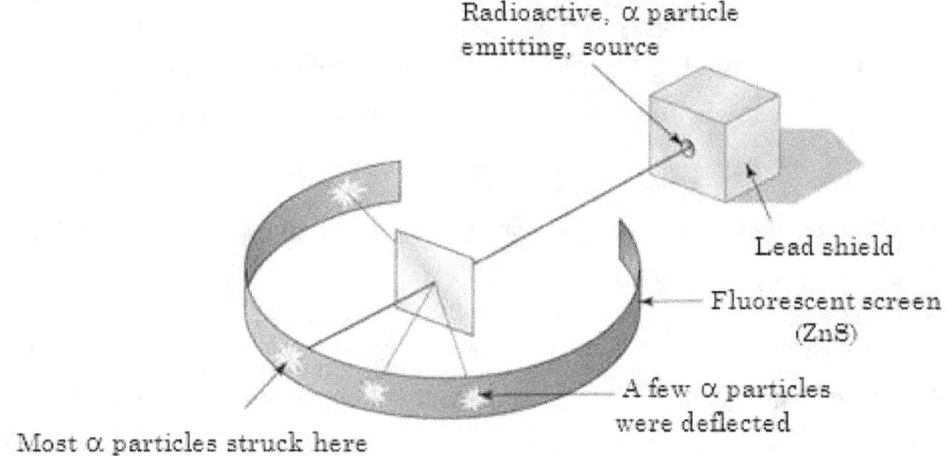

Radioactive, α particle emitting, source

Lead shield

Fluorescent screen (ZnS)

A few α particles were deflected

Most α particles struck here

This led to the extraordinary conclusion that, since the alpha particles had encountered neither proton nor electron, they must have travelled through empty space and this completely contradicted the concept of a completely solid atom which was replaced by the nearly inconceivable conclusion that the atom was essentially empty space.

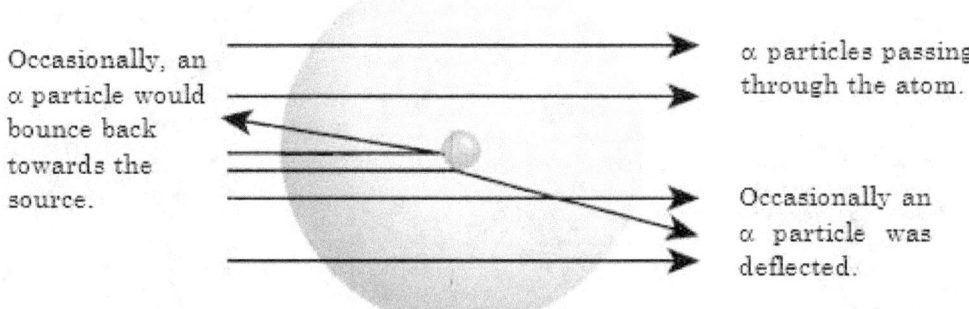

Occasionally, an α particle would bounce back towards the source.

α particles passing through the atom.

Occasionally an α particle was deflected.

There were, however, two further problems to be resolved.

If the atom was almost empty, technically called *free space* in physics, then '**Why did some alpha particles bounce back**?' and '**Where are the protons and electrons as they must be there somewhere**?'

The only conceivable explanation was that there was a very small, positively charged, central part of the atom which was small enough that the atom was indeed still essentially free space. This small centre had to be positively charged since there was no other way the, positively charged, α particles could be repelled.

This answered both questions: most alpha particles passed straight through the free space but some were repelled by collision with the positive nucleus. Therefore the atom must comprise mostly empty space but contain a tiny, positively charged, nucleus.

Lord Rutherford was fascinated by the nucleus and radioactivity but was not particularly interested in the electron and concluded that the electrons simply existed outside the nucleus and circled it almost like a train of electrons.

On the basis of this dsicovery, the Japanese physicist, Nagaoka Hantaro (pictured right on a commemorative, 2003, stamp) rejected Thomson's *plum pudding model* and proposed his *Saturnian Model* which, essentially proposed that the atom comprised a small, but very dense, nucleus surrounded by a circle of travelling electrons and was inspired by the planetary model of our solar system.

The nucleus comprised most of the mass of the atom and was surrounded by a revolving set of electrons bound to the nucleus by electrostatic attraction.

Nagoaka was half right and half wrong and this new concept of atomic structure was revolutionary and led to a surge of interest and experimentation since the physicists' modest belief that nothing could be discovered about the atom was now clearly incorrect.

It also produced what physicists enjoy most which is the opportunity to investigate and solve questions which had never previously been conceived.

More problems arose however since, according to classical electrodynamics, Rutherford's and Nagoaka's concept of electrons simply circling the small positively charged nucleus could not be correct for three reasons.

1. What was to stop the electrons simply being pulled into the nucleus by the positively charged protons?

2. A charge circling another charge would also spiral into the nucleus emitting radiation as it did so:

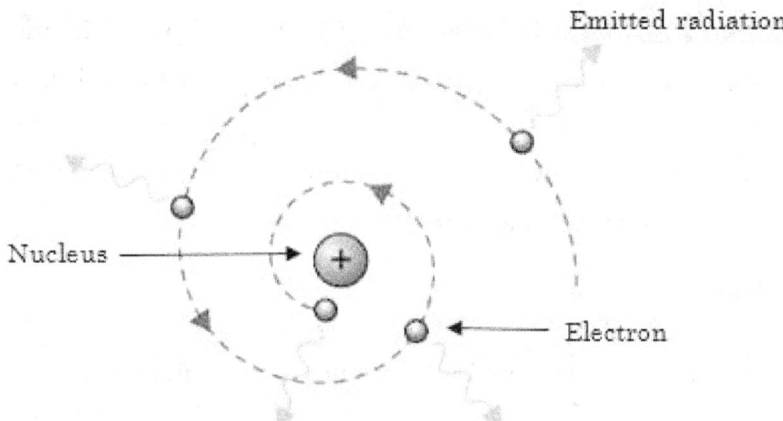

Neither of these events were observed and it also controverted the observation that moving electrons radiate energy which is, to this day, the reason why radio and tv aerials work.

3. The electrons revolving around the nucleus would repel each other and there was no explanation for the attractive power of the, positively charged, nucleus overcoming the mutual repulsion of the, negatively charged, electrons.

It also did not explain a very important analytical technique, the, qualitative, *flame test*. The main metals determined, essentially by lobbing some of the material into a flame and observing the newly produced colours are as follows:

Metal	Flame Colour	Metal	Flame Colour
Lithium	Red	Barium	Green
Sodium	Yellow	Copper	Blue – green
Potassium	Lilac	Strontium	Crimson
Calcium	Brick red	Tin	Blue – white

This could not be explained by any theory to this point but it inspired Niels Bohr's model of the atom.

The Rutherford model could not begin to explain this observation but it did lead to investigation by Niels Bohr (pictured left) whose conception of electrons existing in different levels i.e. different distances from the nucleus did explain that but still did not explain why the electrons did not spiral into the nucleus which is discussed later. Nevertheless Bohr's theory was incredibly important leading to the theory of quantum mechanics and, for our purposes, nuclear magnetic resonance.

All of this is discussed later.

The next matter to consider is F.W. Aston's discovery of the existence of isotopes through his invention of mass spectrometry.

Mass Spectrometry and the Discovery of Isotopes

In 1911, soon after Rutherford, Geiger and Marsden had discovered the existence of the nucleus, F.W. Aston (pictured right), a protégé of J. J. Thomson was invited to join him at the Cavendish Laboratory at Cambridge University to further investigate the *anode rays* which had been discovered by Eugen Goldstein in 1886. These were intriguing since the deflection of a single beam of cathode rays, electrons, produced a deflected beam but only one beam indicating that cathode rays were all exactly the same. In contrast, the anode rays were deflected by various amounts by the magnetic field indicating that the anode rays were **not** all the same.

There were still three problematic questions.

1. If the atom contained a positive nucleus and all nuclei of a particular element were identical then the deflection of all anode rays should produce one single deflected beam. This did not occur and in a magnetic field more than one deflected anode ray was detected.

2. The charges of the proton and the electron were both now known to be equal and opposite but adding up the masses did not explain the known masses of atoms which were inevitably greater than that predicted. This was before Moseley introduced the concept of atomic number and its significance had to wait several years. Likewise the existence of the neutron, the neutral particle with almost exactly the same mass of a proton, though its existence was conceived, had yet to be proven.

3. The relative atomic masses of all known atoms had been measured by other methods. Setting one atomic mass unit as that of the lightest, hydrogen, and ignoring the mass of the electron since being so small it would disappear in the rounding up or down of the mass, all other atoms should be whole number multiples of the mass of hydrogen. No elements had relative atomic masses that were exact multiples of the relative mass of a hydrogen atom.

Aston was intrigued by the relative atomic mass of, the relatively recently discovered element, neon and the relative atomic mass of the chlorine molecule.

Compared to hydrogen, whose relative mass was set at 1 atomic mass units (amu), or even relative to oxygen whose relative mass had been measured to be sixteen times greater than that of a hydrogen atom, the relative atomic mass of neon was 20.2 amu.

This was inexplicable, as was the relative atomic mass of the chlorine atom (35.5 amu) and that of the chlorine molecule, Cl_2, (71 amu) and it was Aston's genius to conceive of the idea of chemically identical atoms of the same element but of different mass and also to design an experiment to establish the truth or falsity of this idea.

These two achievements opened a whole new line of investigation into the structure of the atom and also a new way of analysing the composition of organic molecules.

The importance of these achievements cannot be overstated.

Aston knew that positive ions could be accelerated when passing through an electric field and could be deflected by a magnetic field and it was his inspiration to combine the two fields in succession.

However, having conceived of a system where ions could be accelerated by an electric field and deflected according to their mass/charge (m/z) ratio there were still practical difficulties to be overcome: the atoms had to be ionised, the entire apparatus must operate in a vacuum otherwise the ions would be absorbed by the other gases present and the ions had to be detected. In addition the entire glassware apparatus had to be constructed.

There were no issues with constructing an electric field since that was created by simply connecting two metal plates to an electricity supply (battery) and the magnets were easily obtained. The beam was focused by passage through a negatively charged plate with a central hole. Likewise there was no problem with the glassware as Aston, since his student days, had been a keen glassblower and blew his own apparatus.

The element was vaporised by simple heating and ionised by glow discharge (a plasma formed by passage of an electric current through the gas) whilst the pressure within the apparatus was reduced by a vacuum pump.

Finally the ions were detected by the exposure of a calibrated photographic plate which was developed according to, by then, well established photographic processes. Since the ions were individually deposited on the photographic plate the intensity of the peaks was measured by measuring the height of the deposits with a Vernier scale.

Although modern instrumentation is more sophisticated with peaks now being detected electrically, even lower vacuum pressures and more precise measurements and it is also computer controlled, the essential principles remain unchanged from Aston's first mass spectrograph which is illustrated below.

Aston's mass spectrograph

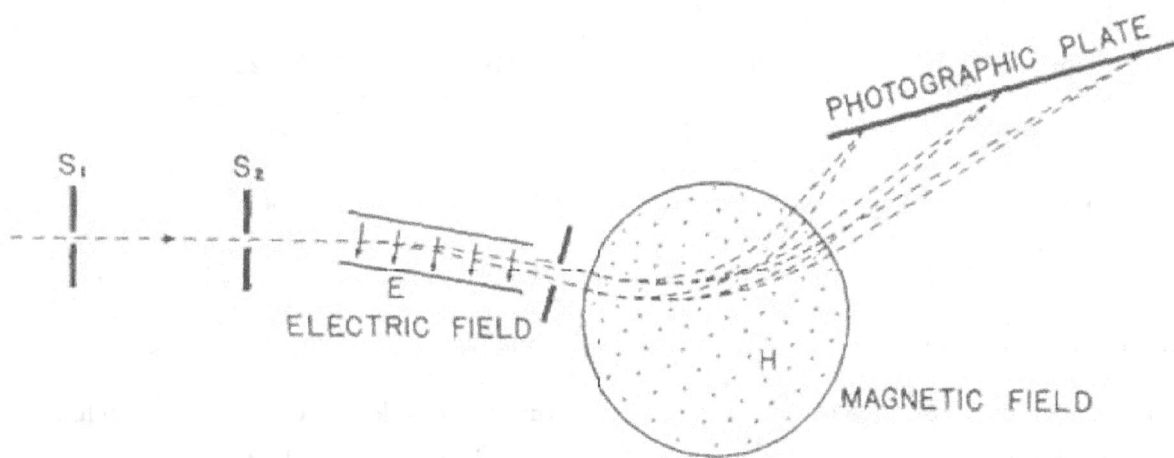

Aston's spectrograph was a great success and demonstrated for the first time the existence of isotopes, chemically identical atoms with the same number of protons but differing numbers of neutrons.

Two examples demonstrate its importance: the relative atomic masses of neon and chlorine, as discussed below.

Neon

Aston demonstrated that the relative atomic mass of neon was 20.2 atomic mass units (amu) because it is comprised of three isotopes of relative atom mass 20, 21 and 22 in the relative proportions: 90.92%, 0.26% 8.82% as shown.

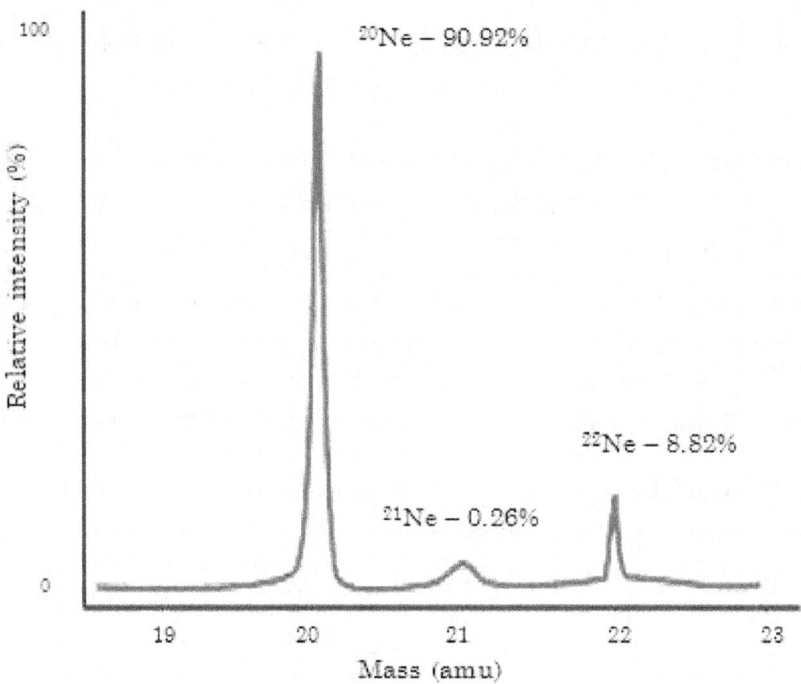

The relative atomic mass is calculated by multiplying the mass of each atom by its abundance and summing the totals. To calculate the relative abundance is straightforward: sum all the peak heights and then divide the height of each peak by the total.

For neon the total abundance of the peaks is 114 + 0.2 + 11.2 = 125.4 and so the relative abundances of the three peaks are as follows:

Mass/charge:	20	21	22	
Peak height:	114.00	0.32	11.06	
Relative peak height:	114.00/125.4	0.32/125.4	11.06/125.4	
=	90.92	0.26	8.82	%

The relative peak height displayed above is the relative abundance of the ions.

To calculate the contribution each of these ions makes to the overall relative atomic mass is achieved simply by multiplying the relative abundance of each peak by its mass and summing the three results e.g.

$$RAM_{Ne} = (20 \times 90.92/100) + (21 \times 0.26/100) + (22 \times 8.82/100)$$

$$= 18.18 + 0.06 + 1.94 = 20.18 \text{ amu}$$

So with rounding the relative atomic mass of neon is measured to be 20.2 is explained.

Chlorine

The spectrum of neon, above, is displayed as it would have been drawn as a graph on the basis of the measurements of the spots on the photographic plate. The, measured, relative atomic mass of the chlorine atom was also a source of mystery as it is 35.5 amu. Aston discovered two isotopes of chlorine ^{35}Cl and ^{37}Cl in the proportion 3:1 as shown below and which is portrayed in the modern, conventional, format.

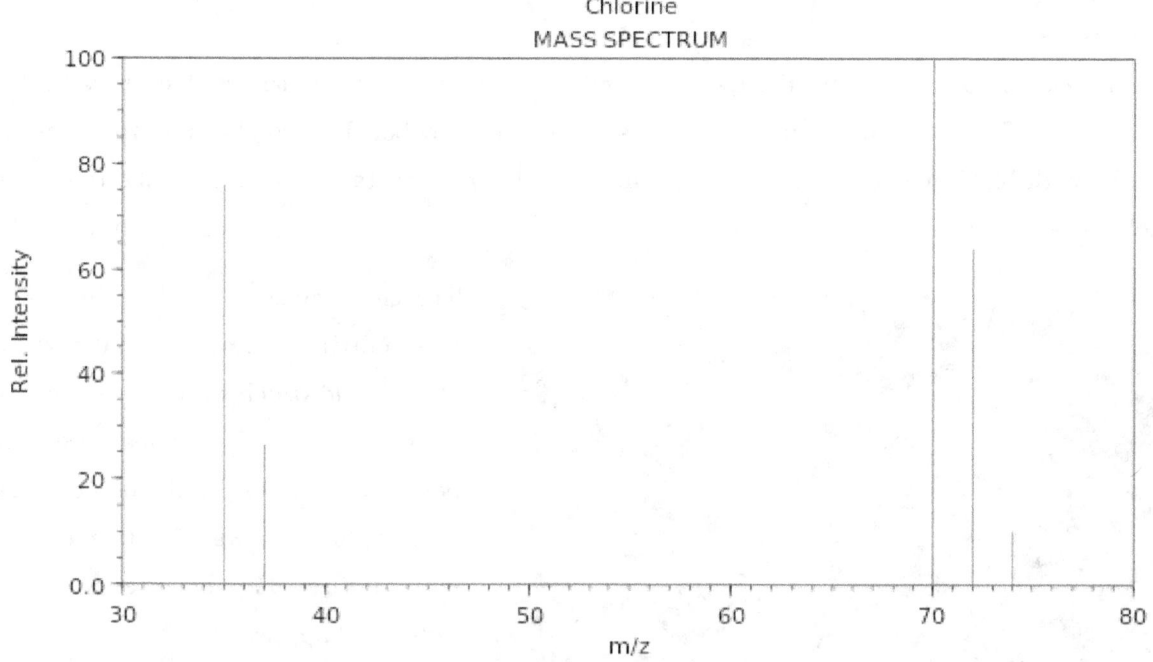

NIST Chemistry WebBook (https://webbook.nist.gov/chemistry)

The relative atomic mass (A_r) of chlorine is measured and calculated as follows:

$$\mathbf{RAM_{Cl}} = (35 \times 75/100) + (37 \times 25/100) = 26.25 + 9.25 = \mathbf{35.5\ amu}$$

This also explained why the relative molecular mass of the chlorine (Cl_2) is 71 since gaseous chlorine consists of the following molecules: $^{35}Cl^{35}Cl$; $^{35}Cl^{37}Cl$; $^{37}Cl^{37}Cl$ in the proportions 9:6:1 meaning that the relative molecular mass may be calculated as follows:

$$\mathbf{M_r(Cl_2)} = (70 \times 9/16) + (71 \times 6/16) + (72 \times 1/16)$$

$$= 39.38 + 26.63 + 4.5 = 70.5 \approx \mathbf{71\ amu}$$

The explanation for the mass of the atom exceeding the mass of the protons and electrons contained within it was given in 1932 by James Chadwick with his discovery of a neutral particle of similar mass to the proton. The significance of the proton (atomic) number is discussed next followed by a discussion of the discovery of the neutron which, although occurring many years apart, answered the following two questions which are of significance for this volume:

1. Is the number of protons of any significance or just an interesting observation of no consequence?
2. Why are the relative masses of atoms greater than the sum of the relative masses of the protons and electrons?

Moseley's Determination of the Significance of the Atomic Number

Is the number of protons of any significance or just an interesting observation of no consequence?

The significance of the number of protons was discovered in 1913 by Henry Moseley (pictured below in 1910), who was tragically killed at Gallipoli, but whose inspirations resolved the issues about Mendeleev's Periodic Table.

X-rays had been discovered by Wilhelm Röntgen. In 1895, whilst investigating the cathode rays that had been discovered by J.J. Thomson, Röntgen noticed crystals on a nearby bench were glowing. He covered his cathode ray tube with black paper and when he again applied a high voltage across the tube the crystals fluoresced, i.e. glowed again.

The clear conclusion was that the tube was emitting a previously unknown type of light and which were rays which could pass through most substances, only being blocked by thick pieces of lead. This explosive news, of course, led to the medical x-ray diagnostics and numerous industrial applications.

Indeed, one of Röntgen's first papers included an x-ray image of the hands of his wife, Bertha.

This invention caught the public imagination and led to some bizarre inventions such as a machine for determining exact feet measurements with many shoe shops advertising their x-ray machine for precise shoe fittings and, even more extraordinarily, radioactive toothpaste which, a by-product of the production of mantles for gas lamps, was advertised as imparting health benefits, including antibacterial action and contributing to the strengthening the defences of teeth and gums.

The production of x-rays for experimentation and industrial application was hampered however until 1913, since x-ray tubes were prone to breakage under the high voltages necessary for scientific application. This was solved by the invention of the Coolidge x-ray tube which could withstand voltages of up to 100 kV and was reliable as the vacuum was extremely low.

The principle was the same but with electrons emitted now and not absorbed on their journey, they were a reliable source of x-rays.

Moseley was fascinated by the invention and switched his studies to the interaction of x-rays and metals. He discovered that each element produced a unique x-ray spectrum and this discovery in itself led to the the entire discipline of x-ray spectroscopy which has been used for numerous applications from the detection and quantification of tiny quantities of impurities in metals through to the study of the interaction of molecules with metallic surfaces. A schematic diagram of the apparatus is shown below:

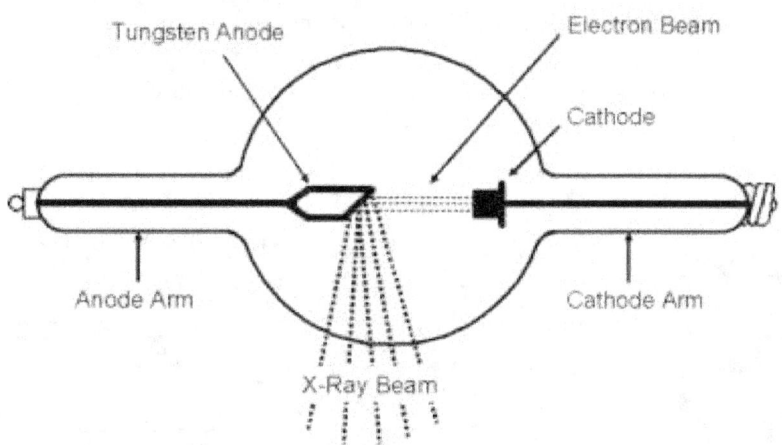

With the encouragement of William Henry and William Lawrence Bragg, father and son who uniquely shared the Nobel Prize for Physics in 1915 for their invention of the analytical technique of x-ray crystal diffraction, Moseley measured the x-ray spectra of thirty nine different elements.

William Henry Bragg noted that Moseley had missed a couple of peaks in the x-ray spectrum of platinum, and guided Moseley to realise that the x-ray spectrum was characteristic of the individual metal. After attempting numerous correlations, Moseley determined that the square root of the frequency of particular peaks was directly proportional to the number of protons in the nucleus as shown overleaf in the graph taken from Moseley's original paper. Even more significantly, Moseley grouped the lines into groups denoted K, L and M lines and, working with Niels Bohr's exposition of the electronic distribution of atoms, he was able to differentiate between the groups and assign some lines as arising due to transitions from inner to outer shells.

Again, it is impossible to underestimate Moseley's discovery who resolved three issues:

a. He identified that the number of protons was a fundamentally significant property and created the concept of the *atomic number*.

b. His work corresponded with the calculations and insights of Niels Bohr (discussed later) into electronic arrangements.

c. Moseley also explained some discrepancies in Mendeleev's Periodic table of the Elements. Mendeleev, whose name is known by all chemistry students for his periodic arrangements of the elements, was not the first to try to create patterns but he was the first to leave gaps hence predicting the existence and properties of as then unknown elements. However, Mendeleev himself made a couple of deliberate mistakes. He had assembled the elements in order of atomic weights (as the masses were then termed) but ignored this when convenient as, for example, with cobalt and nickel. Mendeleev assigned these two metals in that order on the basis of their properties and ignored the fact that cobalt, being slightly heavier than nickel, should have been placed after nickel.

Moseley's work resolved this since by placing the elements in order of the number of protons, *atomic number*, he reconciled the chemical properties with the element's atomic number and further demonstrated that the number of protons, **atomic number**, is *the* fundamental property of an atom.

With these discoveries, Henry Moseley became world famous almost overnight but his death at Gallipoli, at the age of twenty six, robbed the world of untold future discoveries and his demise led to the rule that scientists of great esteem and achievement should not be permitted to serve in places of danger.

The graph of proton number versus square root of frequency in Moseley's first published paper on atomic number is reproduced below.

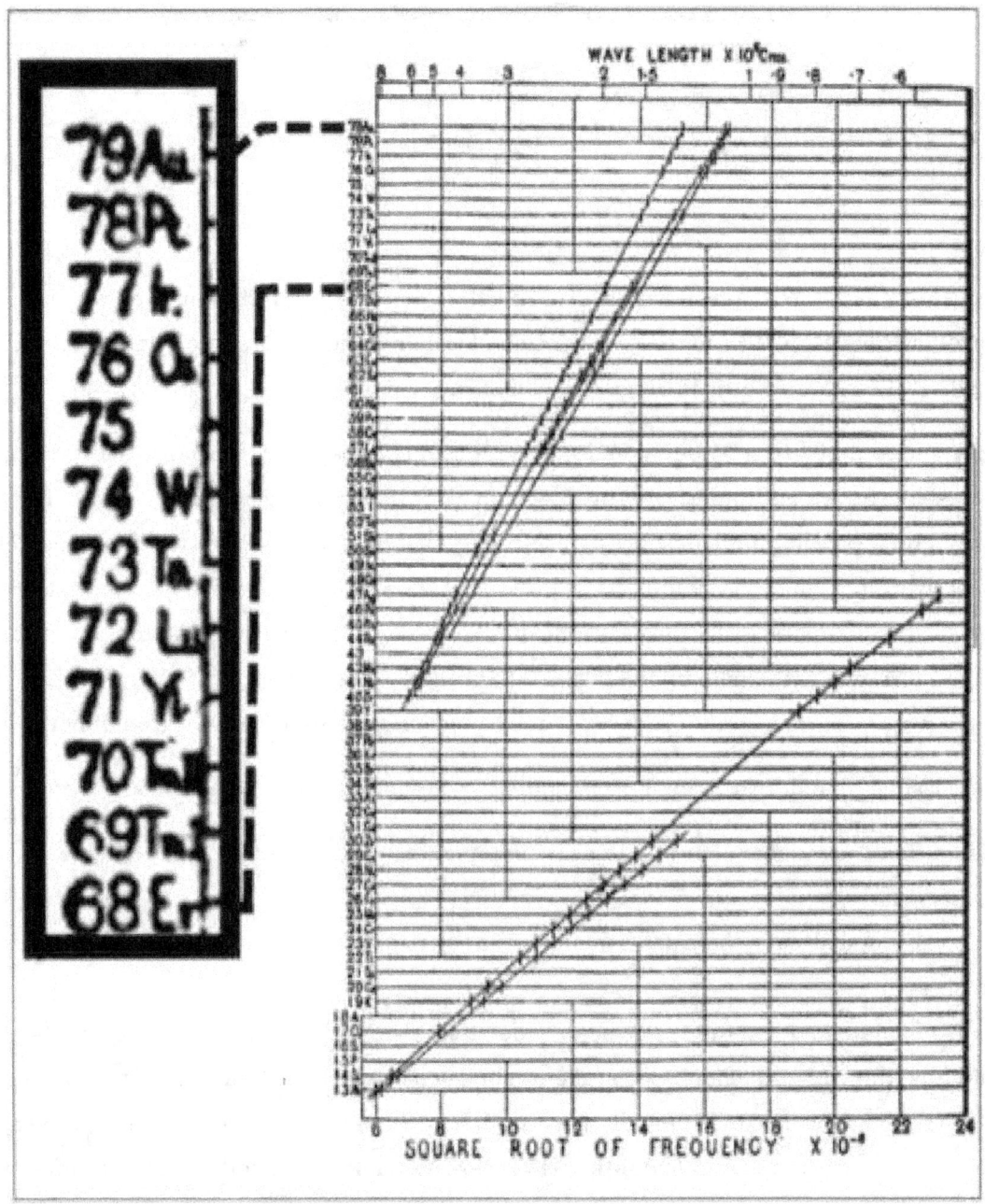

We can now also answer the second question:

Why are the masses greater than the sum of the masses of the protons and electrons?

22

The Discovery of the Neutron

James Chadwick's most elegant experiment answered the question:

Why are the masses greater than the sum of the masses of the protons and electrons?

Writing this in April 2020, one conundrum is analogous to the current concept of dark matter and neutrinos. Dark matter, still a theoretical explanation for the apparently missing, 95%, mass of the universe appears impossible to detect and it is the same with neutrinos. Neither dark matter nor neutrinos appear, *to us*, to interact, with the world. If they exist then they must do so even if just passing through and in some way we have yet to conceive.

The early 20th century problem was how to explain why a neutral atom containing equal numbers of positively charged protons and negatively charged electrons is more massive than the summed masses of the protons and electrons. It was clear that the remaining mass must be due to an uncharged particle. This particle would ignore and would not be accelerated by an electric field and would pass through a magnetic field as if the field did not exist. It must exist but how to detect it?

 The answer was a beautifully simple and ingenious experiment designed by Sir James Chadwick (pictured left) in 1932. Chadwick was yet another former student of Lord Rutherford, and for his discovery of the neutron he received the Nobel prize for Physics, an almost unheard of, three years later.

Many physicists, including James Chadwick, reasoned that a neutral atom that contained equal numbers of of positive particles, *protons*, and negatively charged particles, *electrons*, but whose mass exceeds the total mass of those particles must contain some other, uncharged, particles. Since the mass of the electron was 1/1840th the mass of a proton, it would either contain 1840 neutral particles for every proton or there must be another particle. That particle, if of roughly the same mass as a proton, would only require a handful rather than thousands of neutral particles and seemed more sensible and realistic. The question was: ***How can we detect a particle which is oblivious to both electric and magnetic fields and appears to not interact at all with any other known substances***?

The solution was brilliantly simple and extremely elegant.

Chadwick reasoned, as described above, that it was highly unlikely that the missing mass was due to tens of thousands of small particles. He was also aware that one of the emissions of radioactive materials were the equivalent of helium nuclei: a doubly charged ion but with a relative mass of four indicating that these particles contained these unknown particles as well as protons.

$$^{4}_{2}He^{2+}$$

It is often described as *the* helium nucleus. Some find that confusing since there has been no helium present. In reality, the α-particle is merely the **same** as a helium nucleus.

Chadwick hypothesised that the hidden mass of the nucleus was due to particles of the same mass of a proton since that reasoning required the existence of only two neutral particles of the same mass as the neutron. If they had the same mass as the electron then there would need to be 3,280 of them. Two particles seemed arther more realistic!

His elegant experiment was to direct the beam of radioactive emissions towards a beryllium film in an evacuated system. Classical, Newtonian, mechanics calculations suggested that the momentum of the emitted alpha particles would be sufficient to eject particles of similar size to the proton. The problem, again, was how to detect the neutral particles which were unaffected by electric or magnetic fields.

Again, the solution was derived from classical mechanics: the emitted neutral particles would, in turn, due to the conservation of momentum, eject protons from material upon which it impinged. Due to their charge, the protons could then be detected and the paths taken also enabled an assessment of the mass of the neutral particle. Chadwick selected a thin film of paraffin wax as illustrated below and the experiment was triumphantly successful since protons ejected from the beryllium target were detected.

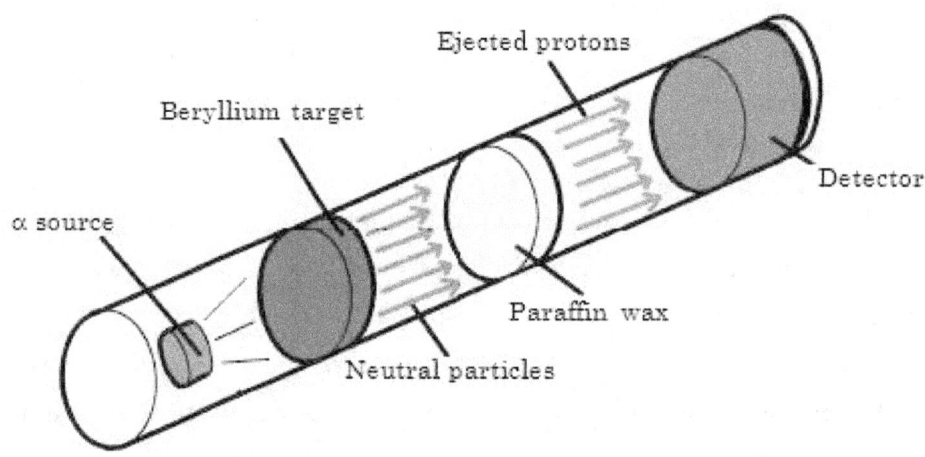

The conclusion was that the neutral particle was of approximately the same mass as a neutron and explained many observations quite perfectly. For example, the α-particle was now known to constitute of two protons and two neutrons and explained the existence of all isotopes whose existence had been discovered almost twenty years earlier by F. W. Aston.

We can turn to the arrangement of electrons around the nucleus and Niels Bohr's model of the hydrogen atom.

Chapter III

The Electronic Distribution Within The Atom

The Bohr model of the atom

This chapter considers:-

- ▪ Emission spectra
- ▪ The wave – particle duality of light and the photoelectric effect
- ▪ Niels Bohr's theory of the structure of the hydrogen atom

all of which led on to de Broglie's conception of wave – particle duality, the Schrödinger wave equation and the arrangement of electrons in the atom (Chapter IV) which utterly transformed understanding of ionic and covalent compounds and bonding, leading to innumerable chemical discoveries which are addressed in Chapter V.

Many physicists, including Lord Rutherford himself, were dissatisfied with the concept of electrons in one circle travelling around the nucleus but, as mentioned previously, Rutherford had other interests.

Niels Bohr rejected the concept on the grounds of electrostatic repulsion and, perhaps inspired by the planetary model of the solar system, suggested that, rather than one revolving circle of mutually repelling electrons, there were actually a number of rings of electrons revolving around the nucleus.

There was also the problem of reconciling the structure of the atom with the hydrogen emission spectrum and the colours of many metals in a flame. With the invention of the spectroscope the actual wavelengths of the emitted light could be measured extremely accurately.

There is little that a chemist enjoys more than finding patterns between substances, as in the Periodic Table, and the concept of functional groups facilitated the study of organic chemistry whilst physicists love finding relationships between data with an ensuing mathematical formula.

The study of the emission spectra satisfied both desires and this led to Bohr's model of the atom which led on to our current understanding of the structure of the atom and these are now discussed.

Emission Spectra

For centuries, it had been known that white light is comprised of the colours of the rainbow. Newton had demonstrated this by splitting white light with a prism and then recombining the coloured light to form a beam of white light.

In 1814, a lens maker, Josef von Fraunhofer (pictured right) extended Newton's work by spreading the dispersed light even more widely and demonstrated that sunlight did not comprise a continuum but that there were actually dark lines in the spectrum. Fraunhofer was able to accurately map and measure the wavelength of 574 of such lines. The origin of these lines was, however, a complete mystery

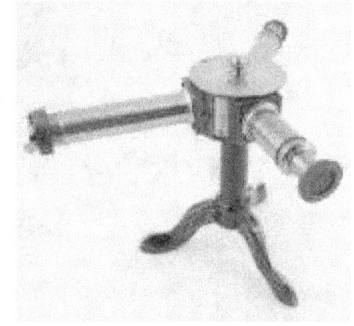

A very major step forward occurred when Gustav Kirchhoff (pictured left) invented his laboratory spectroscope (pictured right). Kirchhoff, working with Robert Bunsen (below left), demonstrated that if he illuminated the flame with light from one of Bunsen's burners, the light turned into dark shadows. Kirchhoff managed to combine the light from the sun and from a flame together and the dark lines from both superimposed on each other exactly.

This was a colossal achievement not just in terms of the experimentation but the conclusion was ground breaking in many ways. It showed that emission and absorption of light were complementary and effectively the opposite of each other and that the sun was made of the same types of atom as found on Earth.

The consequences continued.

Kirchhoff and Bunsen tested many metals and solid compounds simply by holding them on a spatula in the flame and measured the emitted wavelengths of the bright light. The ability to measure precisely the wavelengths of star light permitted Edwin Hubble (1928) to demonstrate, from the *red shift* of hydrogen, that the stars are retreating from us and speeding up as they travel.

The ramifications for chemistry and geochemistry of the discoveries of Kirchhoff and Bunsen were immense. Microgram quantities of metals could be detected in a flame leading to huge advances in geology and geochemistry. Numerous chemists mapped the spectral lines of all known elements. Other lines which could not be assigned to known elements led directly to the discovery of the elements rubidium and caesium which were named after the most prominent red and blue lines in their spectra. Rubidium has a very similar spectrum to that of potassium and it was only through very precise measurements that rubidium could be distinguished and hence discovered.

There was still, however, no explanation for the appearance of the lines, their wavelength or the fact that the lines coalesced.

In 1885, Johann Balmer (pictured right) studied the measured wavelengths of hydrogen's visible spectrum which is shown below.

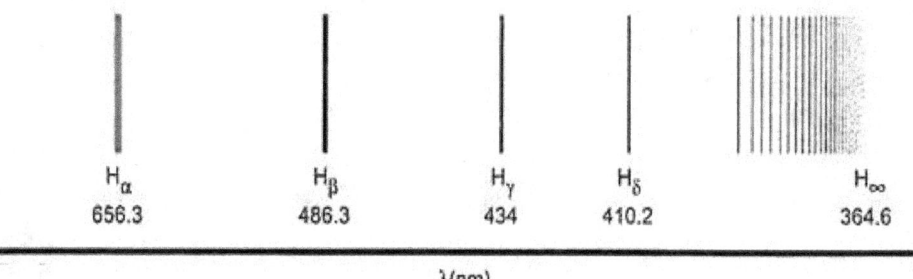

H$_\alpha$ 656.3   H$_\beta$ 486.3   H$_\gamma$ 434   H$_\delta$ 410.2   H$_\infty$ 364.6

λ(nm)

Balmer, derived a formula relating the lines to each other which was:

$$\lambda = B\left(\frac{n^2}{n^2 - 2^2}\right)$$

where λ is the wavelength, B is a constant and n is a integer of the form 1,3,4,5 . . etc but which is never equal to 2 since if n = 2 then the denominator becomes zero and the quantity in bracket equates to infinity.

Another physicist, Johannes Rydberg, determined the relationship between the wavelengths of the lines and derived the following formula:

$$\frac{1}{\lambda} = R_H\left(\frac{1}{2^2} - \frac{1}{n_2^2}\right)$$ which simplifies to $$\frac{1}{\lambda} = R_H\left(\frac{1}{4} - \frac{1}{n_2^2}\right)$$

where λ is the wavelength and n is an integer 1,3,4,5 . . . etc; and R$_H$ =1.097 x 10^7 m^{-1} .

This was a highly significant development but it could not be the final answer since it still did not explain the origin of the lines and it did not work for all elements.

In 1906, the Harvard physicist, Theodore Lyman, was studying the characteristic emissions of hydrogen in the ultraviolet region. He identified three series of lines which he designated α , β and γ.

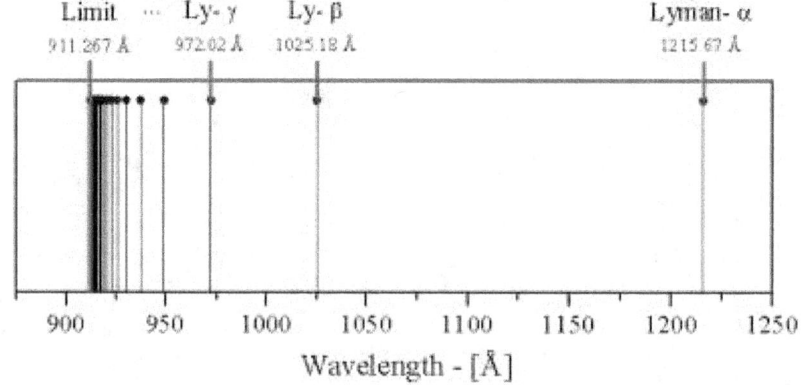

Limit ... Ly- γ   Ly- β   Lyman- α
911.267 Å   972.02 Å   1025.18 Å   1215 67 Å

900   950   1000   1050   1100   1150   1200   1250

Wavelength - [Å]

Lyman noted three significant aspects of the spectrum which resonated with Balmer's discovery of the visible spectrum of the hydrogen atom:

1. There were a series of discrete lines.
2. With decreasing wavelength, the lines came closer together.
3. At a particular wavelength the lines coalesced into a continuum.

After much work on different correlations, Lyman determined that the lines could be calculated according to the following formula:-

$$\frac{1}{\lambda} = R_H \left(\frac{1}{n_1^2} - \frac{1}{n_2^2} \right)$$

where λ is the wavelength, n_1 and n_2 are integers 1,2,3,4 . . . but $n_1 \neq n_2$ and $R_H = 1.097 \times 10^7 \ m^{-1}$

This formula calculated all the lines in the Balmer and Lyman series and the other series which had also been discovered (Paschen, Brackett, Pfund and Humphreys).

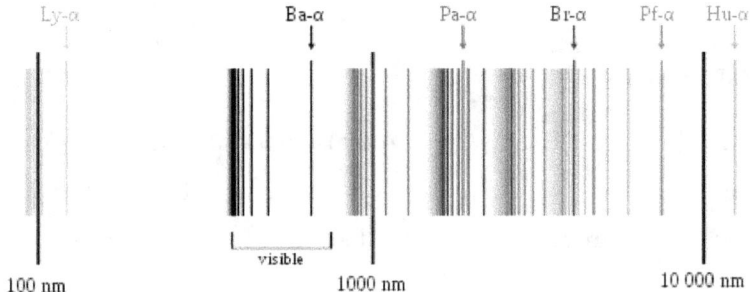

The lines were clearly fundamental and arose from atomic structure since the absorption and emission structures complemented each other perfectly.

To reiterate, the Balmer series lies in the visible region, the Lyman series lies in the ultraviolet region, whilst the Paschen, Brackett, and Pfund series lie in the infrared region.

All lines can be calculated by a formula analogous to the Lyman formula:

$$\frac{1}{\lambda} = R_H \left(\frac{1}{n_1^2} - \frac{1}{n_2^2} \right)$$

where n_1 = 1, 3, 4 or 5 and n_2 = 2, 4, 5 or 6 respectively. The Lyman formula requires that n_1 = 3 and note again that $n_1 \neq n_2$.

This was further, definite, proof of the fundamental relationship of the spectral lines to atomic structure when further was collected from analysis of the spectra of other simple ions with one electron only, specifically the He^+ ion and the Li^{2+} ion. Each of these ions contain only one electron so are analogous, electronically, to the hydrogen atom and their spectral lines follow a similar formula. As an aside, all species with the same number of electrons e.g. $_1H$, $_2He^+$ and $_3Li^+$ are referred to as being *isoelectronic*.

Another hugely significant development was the proposal of Max Planck that energy is quantised i.e. exists as discrete bundles of energy known as *quanta*. Einstein used the concept to explain the photoelectric effect which is discussed in the next chapter.

The Bohr Model of the Structure of the Hydrogen Atom

In 1916, the Danish physicist, Niels Bohr (pictured below right) published a concept of a series of circular orbits indicating possible paths for an electron.

It is a fundamental principle in physics that everything exists in the lowest possible energy state, unless forced to behave otherwise, which explains why a ball dropped down a flight of steps does not stop until the ball is on the ground floor. The ball's gravitational potential energy (GPE) is converted to kinetic energy (KE) which is used to move the ball down the staircase until, on the ground floor, when its GPE is at a minimum. Likewise, a ball is given, acquires, GPE when it is carried upstairs.

Bohr extended this concept to a hydrogen atom.

The single electron exists in the lowest energy level (orbit) unless provided with sufficient energy to move to the next or an even higher energy level. The lowest energy level is the *ground state* whilst all other levels (orbits) which are of higher energy are known as excited states.

Bohr's stepladder analogy to the energy levels of the hydrogen atom accounts for the existence of specific lines in the absorption spectrum and in the emission spectrum.

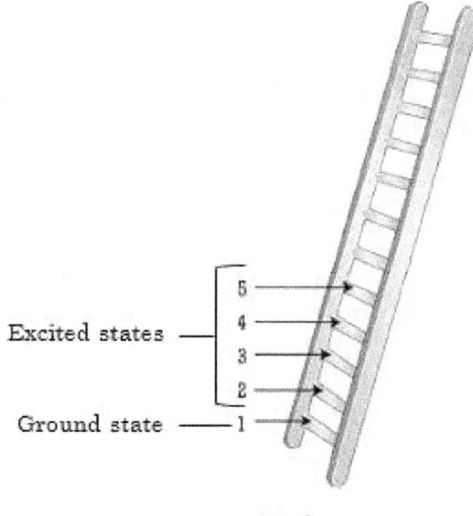

Nucleus

When an atom absorbs radiation it only absorbs specific energies.

The rest of the light travels straight through and the measured spectrum has a number of dark lines, indicating the energies which have been absorbed and hence demonstrating the energies required for the electron to move from the ground state upwards.

When the irradiating energy is removed then the electron relaxes and moves from an excited state to the ground state releasing exactly the same quantity of energy required to move the electron from the ground state upwards. This explains why the absorption and emission spectra are exactly complementary and, superimposed, form a continuous spectrum.

Bohr went further:

1. The stepladder analogy is a start but if the steps, i.e. orbits, are exactly the same distance apart then the wavelengths of the absorbed and emitted radiation would be simpler multiples of the lowest highest energy line (remember that frequency and wavelength multiplied together give us the speed of light – the higher the energy, the higher the frequency and the lower the wavelength).

    They are not simple multiples of the lowest energy (lowest frequency, highest wavelength) line.

    The lines get progressively closer together which implies that the orbits (the steps in the ladder) also get closer together the further one moves from the nucleus.

2. Eventually the lines are so close together that they can no longer be separated and then there are no further lines. This confirms the concept of the orbits getting closer together the further from the nucleus until the lines *coalesce* and can no longer be distinguished.

    Showing only the first four energy levels, the hydrogen atom then, according to Bohr has the structure:

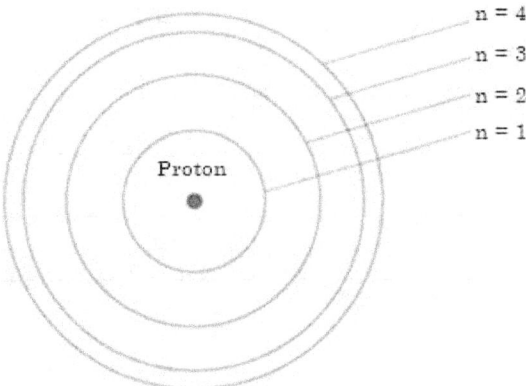

Bohr drew the conclusion that, from his model, the highest energy series of lines (the *Lyman series*) arose from transitions from the ground state (n=1, the orbit closest to the nucleus) and that the next highest energy, *Balmer*, series would be due to absorption of energy from second innermost orbit (n=2).

By assessing the exact wavelengths of the absorption spectrum he was able to calculate the exact distance (in energy terms) of the three next outer levels.

The transitions from n= 2 outwards were determined to be as follows:

Transition	Energy (qualitatively)	Wavelength (nm)	Colour of line
$n_2 \longrightarrow n_3$	Lowest	656	Red
$n_2 \longrightarrow n_4$		486	Aqua
$n_2 \longrightarrow n_5$		434	Blue
$n_2 \longrightarrow n_6$	Highest	410	Violet

3. Bohr was able to explain the existence of all of the remaining series: Paschen, Brackett, and Pfund as follows where n_1, n_2, n_3 etc; are the levels:

The energy differences between n_4, n_5 and n_6 become successively smaller and are observed in the infrared region and are reconciled by Bohr's concept that the further from the nucleus the closer together the energy levels become, as demonstrated above.

Conventionally, it is more appropriate and convenient to discuss the energies of the different orbits as lines as shown below.

This is most easily understood when the two ways of describing the energy levels are considered together with the visible spectrum:

Paschen: $n_3 \longrightarrow n_4 \longrightarrow n_5 \longrightarrow n_6 \longrightarrow n_7 \ldots \longrightarrow \infty$ (ionisation)

Brackett: $n_4 \longrightarrow n_5 \longrightarrow n_6 \longrightarrow n_7 \ldots \longrightarrow \infty$ (ionisation)

Pfund: $n_5 \longrightarrow n_6 \longrightarrow n_7 \ldots \longrightarrow \infty$ (ionisation)

and can be summarised as follows:

Series	n_1	n_2	Spectral lines region
Lyman	1	2,3,4 etc;	Ultraviolet
Balmer	2	3,4,5 etc;	Visible
Paschen	3	4,5,6 etc;	Infrared
Brackett	4	5,6,7 etc;	Infrared
Pfund	5	6,7,8 etc;	Infrared

and visually described as below.

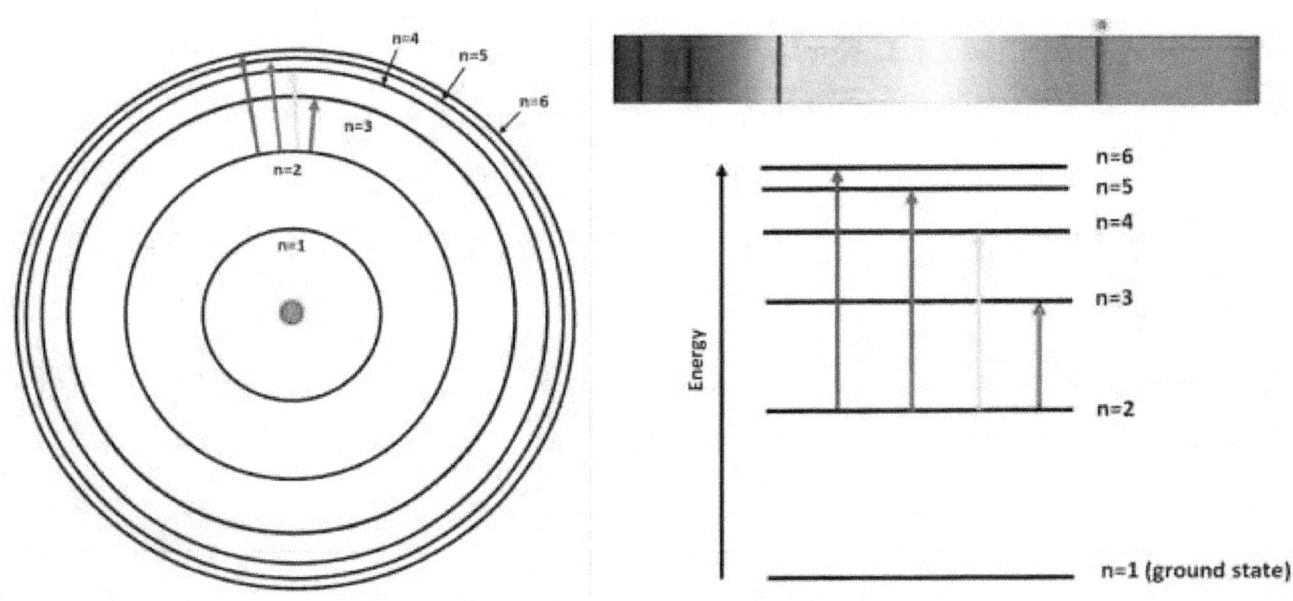

Bohr's model was an almost complete triumph as shown again below:-

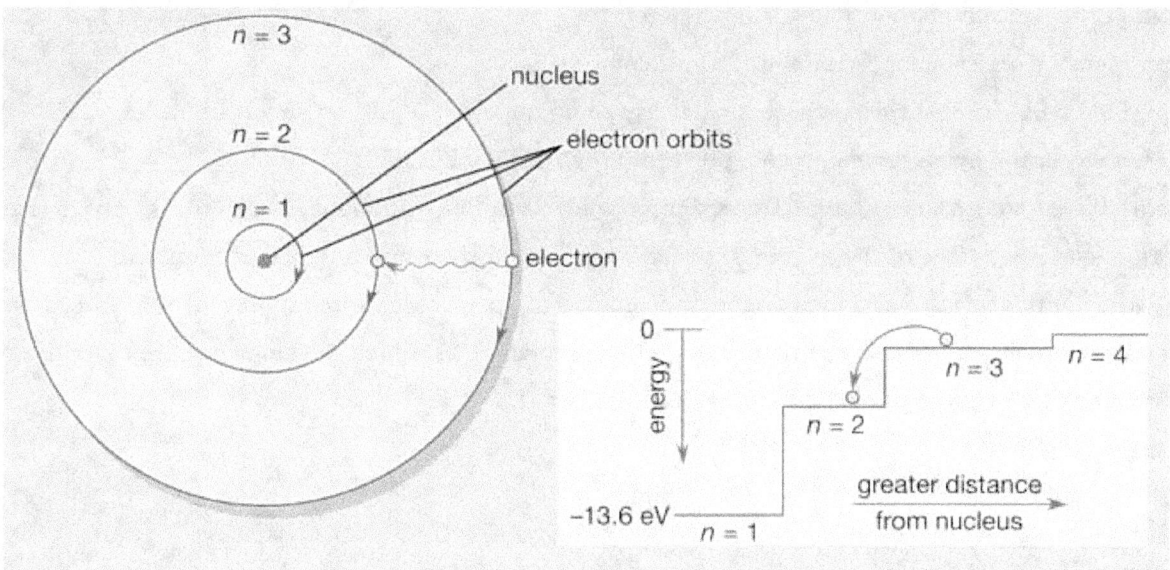

It explained the:-

- Existence of a series of spectra in the uv (*Lyman series*), visible (Balmer series) and the three in the ir regions (*Paschen, Brackett and Pfund series*).
- Reasons for the individual lines within each series.
- Flame test colours.
- Superimposability of the absorption and emission spectra.
- Black lines in the absorption spectra and the coloured lines in the emission spectra which except for the coloured lines is completely black.
- Ionisation since once the electron had acquired sufficient energy to move beyond the highest energy level it was free of the atom which had been ionised.

Bohr's theory was also proven to be a good model since he was also able to explain the spectra of the He^+ ion and the Li^{2+} ions.

Each of the latter two ions involve a single electron orbiting a positively charged nucleus but it proved impossible to extend the model to atoms or ions containing more than one electron since it was impossible to know their velocity and their interactions with other electrons and the nucleus at the same time.

Bohr was able to calculate, using traditional classical electrodynamics of one charge passing by another, in these cases, the, negatively charged, electron orbiting a positively charged nucleus. Once two or more electrons were present, however, the calculations required knowledge of the velocities of both electrons and their interactions both with the nucleus and their mutual repulsions. Knowledge of the velocities of the two electrons was impossible to ascertain and it was at that point that Bohr's model was clearly, a huge step forward, it was not the final explanation.

Moreover, Bohr's theory still did not explain why electrons did not spiral into the nucleus as predicted by classical electrodynamics but the number of answers his theory provided demonstrated a huge step forward in understanding atomic structure.

Bohr's theory:-

* Created the concept of energy levels in the atom;

* Explained the emission spectra of different elements;

* Explained why the absorption spectra perfectly complemented emission spectra but;

* Did not explain why electrons do not spiral into the nucleus but did;

* Explained the ionisation of atoms since when the electron had sufficient energy to rise above the highest energy level it was free of the influence of the nucleus and the atom had become ionised;

* Laid the foundations of the most important chemical spectroscopic technique of all: *nuclear magnetic resonance* which depends on the theory of quantum mechanics which developed approximately ten years after Bohr's theory was published and is discussed next.

Chapter IV

Quantum mechanics

This chapter considers:-

- ■ The behaviour of light (diffraction)
- ■ Einstein's explanation of the photoelectric effect
- ■ The wave – particle duality of the electron
- ■ Schrödinger's wave equation and the electronic distribution within the atom
- ■ The significance of the Noble Gases to electronic configurations.

We examine how Thomas Young demonstrated the wave nature of visible light through his diffraction experiments. Two hundred years later, Max Planck proposed that light was not a continuous stream of energy but that it existed as a stream of discrete particles which became termed *quanta* and Einstein used the concept of the particulate nature of light to explain the *Photoelectric Effect*.

Given that Einstein had proposed light could act as both waves and particles, Louis de Broglie proposed that the reverse must also be true, that particles could behave as waves and, in turn, this led Schrödinger to derive a wave equation explaining the electronic distribution of the atom.

The discovery that electrons exist within specific regions, termed *orbitals*, of the atom led to the explanation of the arrangement of the elements in Mendeleev's Periodic Table, resolved some discrepancies and confirmed G.N. Lewis' suggestion of the significance of the stability of the Noble Gases.

Diffraction

Until the very beginning of the 20th century, physicists considered their discipline to comprise the study of physical objects and of fields.

Essentially all physics can be described as the study of pushes or pulls i.e. *forces* effective physically, by contact, or at a distance, through *fields*. Isaac Newton originally believed that light consisted of particles which he termed '*corpuscles*'. From his own, Newtonian, mechanistic approach that could not be proven and the theory was discarded for a simple reason. If one constructs a barrier with two slits and throws balls at the barrier then the number of balls getting through the barrier and those bouncing off remain the same as the number of balls originally thrown.

In contrast, if a beam of light is shown at the barrier and passes through both slits, two bright spots of light appear on the other side of the barrier but, in addition, the light rays interfere with each other and produce an image of lighter spots and completely dark regions. Thomas Young (pictured right) performed this experiment in 1800 and demonstrated that the light rays must interfere.

If they *interfere constructively*, then they combine, reinforce each other and produce the bright lines (brighter than the impinging light) whilst with *destructive interference* they combine and cancel each other out creating the dark lines. This could not occur if the irradiating light was completely comprised of particles, however small.

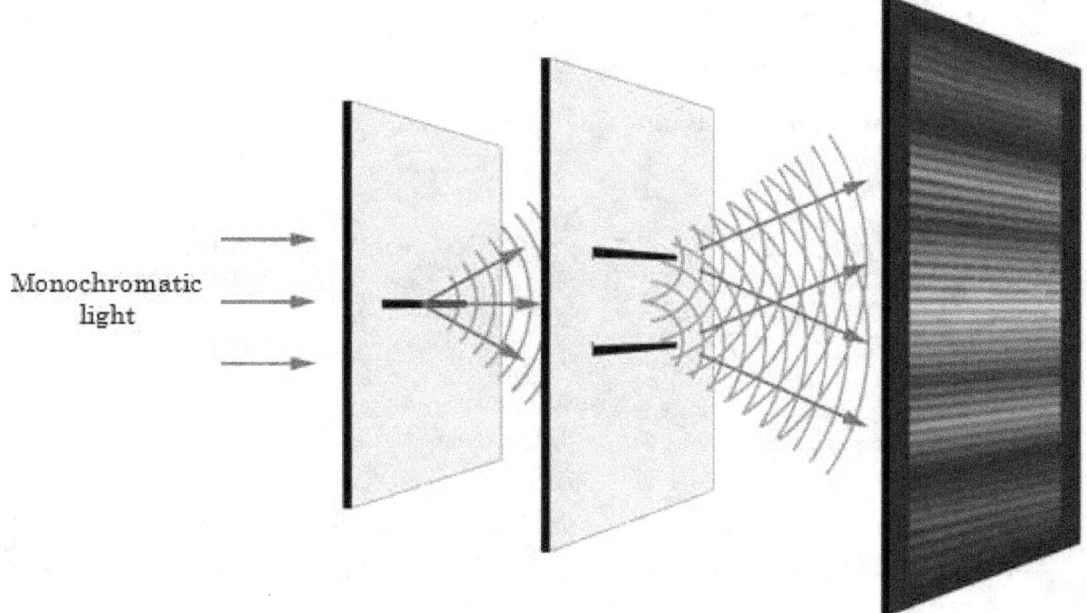

Monochromatic light

There is no way to explain this if light is comprised of particles and the observation was analogous to the effect of two ripples in water which reinforce in places, creating higher waves, and, in other places, cancel each other out creating regions of completely still and flat water.

Huge advances were made over the next two centuries in the theory of waves, leading eventually to the explanation by James Clerk Maxwell (pictured left) that light is formed by the propagation of orthogonal electric and magnetic fields, that is fields at right angles, $90°$, to each other. Numerous experiments confirmed this theory and led to, perhaps, the most dramatic demonstration of the transmission of energy by waves: radio.

This all appeared to demonstrate conclusively that light has the properties of waves and not particles.

Perhaps not.

In 1900, Max Planck proposed the quantisation of energy, using the idea simply as a mathematical trick to explain atomic emission spectra. In essence, he concluded that if the spectra were not continuous i.e. comprised of every single possible wavelength then energy must exist as packets, discrete quantities.

The Photoelectric Effect

In 1905, Albert Einstein used the theory of the quantisation of energy and the concept of light as particles to explain the *photoelectric effect*. It was for this work, and not for his special or general theories of relativity, that Einstein was awarded the Nobel Prize for Physics in 1921.

The photoelectric effect had mystified physicists for years. Essentially, if a beam of visible light irradiates a metal target then very little happens. If, however, the frequency (energy) of the light is increased then there comes a point at which an electric current is generated as illustrated below. If the energy (frequency) of light increases further then the electrons are ejected with greater velocity.

Low energy radiation	Higher energy radiation	Even higher energy radiation

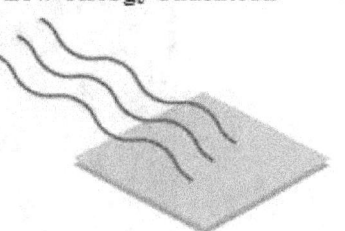 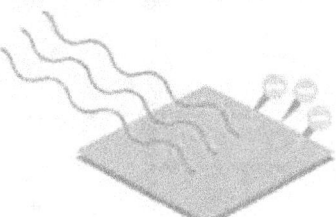 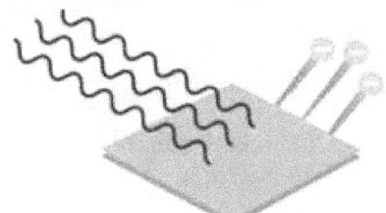

| No electrons
∴ no current | Stream of electrons
∴ a current | A stream of electrons ejected
at high velocity |

 The photoelectric effect was discovered in 1887 by Heinrich Hertz (pictured right) during his research on radio waves. Hertz established that when uv light irradiates a pair of metal plates, the voltage at which a spark occurs varies with the frequency of light. In 1902, Philipp Lenard (left), demonstrated that electrons, then still known as cathode rays, were liberated from the surface of  a metal when irradiated with ultraviolet light.

Visible light had no effect on the metal and the frequency at which a current was detected varied with the metal targeted. This could not be explained but further complications arose when the kinetic energy of the ejected electrons was measured. Once liberated, the velocity i.e. the kinetic energy of the ejected electrons varied with the *frequency* of the light and not with its *intensity* (brightness). Increasing the intensity increased the current generated meaning that an increase in inetist resulted in more elcectrons being ejected.

Electromagnetic radiation can be best described as the means of transferring energy from one location to another without any physical contact between the two locations. This is in contrast to the means by which one particle can transfer energy to another and that can only be through physical contact e.g. collisions. If light was a wave then increasing the intensity (brightness) of even low frequency radiation should allow sufficient energy to be built up by the electron so that, eventually, it could escape from the surface of the metal whatever the frequency. The lower the frequency of light the greater the time that this would take to achieve and even at low frequencies electrons should be ejected, just with a greater time lag but this does **not** occur.

To summarise,

- Once the minimum frequency necessary had been achieved there was no time lag between irradiation and ejection;
- Increasing the frequency did not increase the current i.e. the number of electrons which escaped but did increase their kinetic energy;
- Increasing the intensity whilst maintaining the frequency increased the number of electrons which escaped but only once a minimum frequency of light was employed.

Albert Einstein (pictured right) reasoned that since a certain quantity of energy was required to release the electron this was possible if the irradiating radiation consisted of *particles* not waves. And continued:-

- The kinetic energy of the light particles could be transferred to the electrons. If the energy of the light particles was low then the metal would absorb the radiation as heat. However, once a particular energy was reached this energy would be sufficient to eject the electron from the metal and the electron would be ejected and travel at a low velocity.

This explained the minimum frequency required.

- Increasing the energy of the radiation above this minimum required to allow the electron to escape meant that there was surplus energy. This surplus energy of the light particles could be transferred to the electrons and so the electrons would leave at greater velocity i.e. with greater kinetic energy.

This explained why the velocity of the ejected electrons was proportional to the frequency of the irradiating light, beyond the minimum required.

- Increasing the frequency did *not* increase the current (the number of ejected electrons) but increasing the intensity did achieve precisely that. If the radiation was particulate in nature then the intensity was a measure of the number of particles. Increasing the intensity increases the number of light particles impinging on the metal and increases the number of electrons ejected.

This explained why, at any particular frequency above the minimum required for ejection, increasing the intensity increased the number of ejected electrons i.e. the measured current.

Einstein's theory also explained:-

- **The absence of a time lag:** Provided the impinging radiation had sufficient energy then the electrons would be ejected almost immediately as there would be no need to build up the quantity of energy necessary to eject the electron.

- **The energy needed to eject the electron varies with the nature of the metal:** Different metals held electrons within the structure to varying degrees and so required different energies (frequencies) of electromagnetic radiation. Einstein termed the energy required to eject the electron the metal's *work function* (Φ) and quantified his theory with the deceptively simple formula:

Energy of light = Energy required to eject the electron + kinetic energy of the electron

and, mathematically:

$$hf = \Phi + \tfrac{1}{2} mv^2$$

where: **h** is Planck's constant, **f** is the frequency of the light sufficient to ionise the atom, **φ** is the work function of the metal, **m** is the mass of the electron and **v** is the velocity of the ejected electrons.

In contrast to the wave nature of electromagnetic radiation which explained none of the observations of the photoelectric effect, Einstein's theory explained everything and it was for this that he was awarded the 1905 Nobel Prize in Physics.

Einstein's work was entirely theoretical as he was not an experimental physicist. It remained for others to determine the validity of his theory notably Robert Millikan. Millikan, famous for his elegant, experimental determination of the charge of an electron in 1909, verified the, deceptively simple, equation and demonstrated that Einstein's constant, **h**, was exactly the same as the constant that Planck had produced when he equated energy of light with frequency e.g. $E = h\,f = c / \lambda$

where **E** is the energy of the light, **c** is the velocity of light in a vacuum and λ is the wavelength of the radiation.

Einstein's theoretical explanation of the photoelectric effect became of even more profound significance when it was discovered that as the frequency of the light was increased but not much happened after the first current was detected until, suddenly, a second electric current was detected and then a third and then a fourth etc; These equated to successive ionisations of the original atom and equalled the atomic number, and hence the number of electrons of the target atom.

For example, hydrogen only showed one ionisation (only one electron to be removed) but helium exhibited two ionisations relating to:

$$He \xrightarrow{\;-\,e^-\;} He^+ \xrightarrow{\;-\,e^-\;} He^{2+}$$

whilst lithium displayed three currents:

$$Li \xrightarrow{\;-\,e^-\;} Li^+ \xrightarrow{\;-\,e^-\;} Li^{2+} \xrightarrow{\;-\,e^-\;} Li^{3+}$$

and so on. No atom demonstrated more sudden peaks in the electric current than there were protons in the nucleus and hence the number of peaks also equated to the number of electrons in the original, target, neutral atom.

The Wave – Particle Duality of Matter

The word *profound* has been used a number of times in this volume but the verification of Einstein's explanation of the photoelectric effect was even more than that.

Einstein had explained many matters by suggesting that light was not just a wave but could also be construed as a stream of particles.

This led Louis de Broglie (pictured right) and others to question if the opposite was true:

If a wave can behave like a particle then can a particle behave like a wave?

de Broglie won the Nobel Prize for Physics in 1929 for his answer to this question.

He is probably the only Nobel Laureate to win the accolade based on a doctoral thesis which was also extraordinarily for its brevity, amounting a mere handful of pages.

de Broglie proposed that the momentum of an electron (**p**) was related to its energy by the following relationship:

$$\lambda = \frac{h}{p}$$

where **λ** is the wavelength and **h** is Planck's constant.

Subsequently he wrote that '*The fundamental idea of the thesis was the following:*

The fact that, following Einstein's introduction of photons in light waves, one knew that light contains particles which are concentrations of energy incorporated into the wave, suggests that all particles, like the electron, must be transported by a wave into which it is incorporated.

My essential idea was to extend to all particles the coexistence of waves and particles discovered by Einstein in 1905 in the case of light and photons.

With every particle of matter with mass m and velocity v a real wave must be 'associated', related to the momentum by the equation:

$$\lambda = \frac{h}{p} = \frac{h}{mv} \sqrt{1 - \frac{v^2}{c^2}}$$

where **λ** is the wavelength, **p** is the momentum, **m** and **v** are, respectively, the mass and velocity of the particle whilst **c** is the velocity of light in a vacuum. The equation de Broglie quoted is the so called relativistic version which took account of Einstein's discoveries.

The next problem was to determine whether or not de Broglie's proposal had any validity.

It did.

 The validity of de Broglie's hypothesis was demonstrated at the Western Electric (later Bell) Labs in 1927 by Clinton Davisson (pictured left) and Lester Germer (right) who managed to achieve firing slow moving electrons at a nickel target and obtained a diffraction pattern that would have been expected of a wave. Ironically they were not looking for evidence of the wave nature of the electron but were attempting to study the structure of nickel as
demonstrated below.

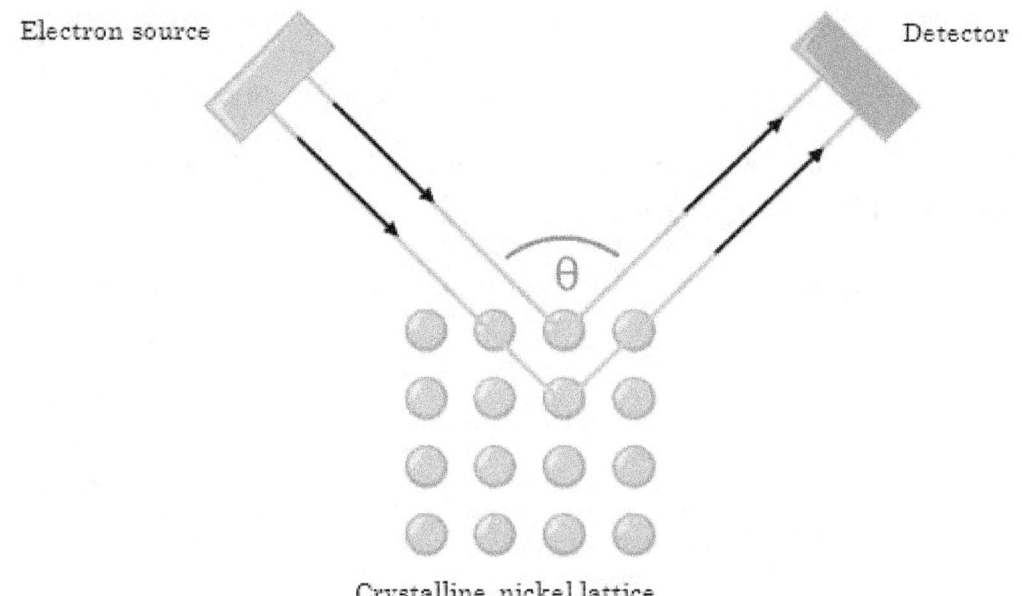

Crystalline nickel lattice

The detector could be moved to measure the electrons at different angles and to their great surprise they discovered that rather than there being a consistent intensity of electrons, there were regular peaks observed at different voltages as shown below:

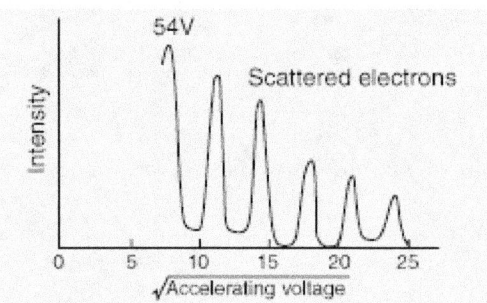

Davisson and Germer resolved this by relating the Bragg angle of diffraction to the de Broglie relationship

$$\frac{1}{\lambda} = \frac{n}{2d\sin\theta} = \frac{p}{h} = \frac{\sqrt{2mE}}{h} = \frac{\sqrt{2meV}}{h}$$

where λ is the wavelength, **n** is an integer, **p** is the momentum of the electron, **m** is the mass of the electron, **V** is the applied voltage, θ is the angle of measurement, **h** is Planck's Constant and **d** is the distance between the atoms in the, crystalline, nickel lattice and **E** is energy.

Nearly simultaneously, the same effect was demonstrated by Sir George Paget 'G.P.' Thomson FRS (below left), the son of Sir J. J. Thomson who had discovered electrons, initially termed cathode rays.

G.P. Thomson was the first person to deliberately search and to find evidence for the wave nature of electrons and in 1927, he discovered revealed the first experimental evidence by studying the diffraction of electrons by a celluloid film and continued to study the phenomena with thin foils of gold and of aluminium and of platinum.

He placed the electron source 1 metre from the target and placed a photographic plate 10 cm behind the target which was 1 μm thick and generated a series of circles.

The experimental set up is shown below:

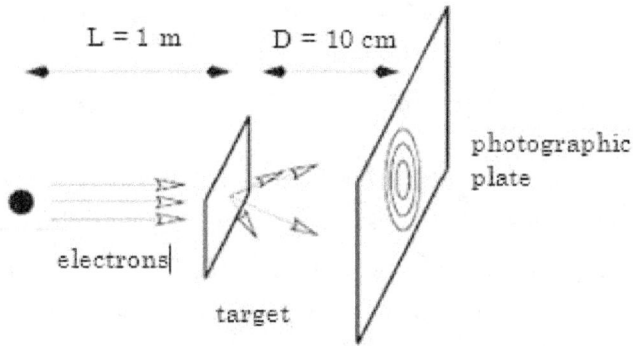

In all cases Thomson discovered and recorded the existence of rings caused by the diffracted electrons whose measurements concurred in full with the de Broglie equation.

The early diffraction patterns recorded in Thomson's first report of his experiment are shown on the right and they are so similar to the diffraction patterns produced by analogous x-ray diffraction experiments on targets of the same material that Thomson's experiment demonstrated beyond doubt that electrons possess wavelike properties.

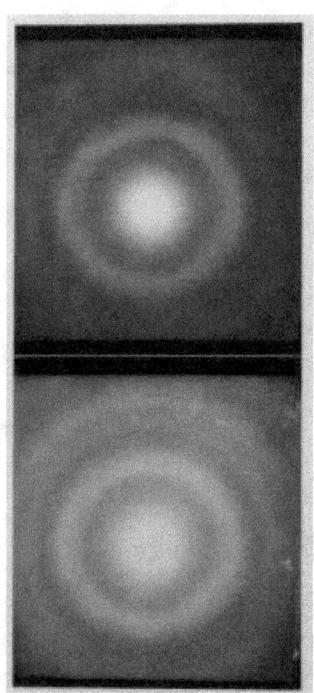

It is noteworthy that J.J. Thomson received the 1906 Nobel Prize in Physics for his discovery of the electron which he regarded as a particle whilst his son, G. P. Thomson, received the 1937 Nobel Prize in Physics for discovering its wavelike properties which he shared with Davisson.

Taking stock of these discoveries:

The period, 1885 – 1932, enveloped numerous discoveries which shook the world of classical physics to its core.

From the modest assumption of Victorian physicists that some questions were unanswerable and that the atom was a discrete solid and indivisible ball about which nothing could be determined, it had been established that:-

- Atoms contain negatively charged species, originally termed cathode rays but now called electrons.
- Atoms contain positively charged species (protons) and the same atoms contain an equal number of protons and negatively charged electrons but the electrons are of far less mass.
- The vast bulk of the density of the atom is contained within the central positively charged nucleus.
- Most of an atom is empty, so called *free space*, essentially a perfect vacuum.
- The electrons exist in a number of orbits circling the nucleus.
- The newly discovered particles, *electrons*, behave as particles and waves depending on the circumstances. Generally, however, and for simplicity, chemists visualise electrons as particles whilst always conscious of their wavelike properties as well.

It was a stunning change and is demonstrated, most extraordinarily, that if one raises an arm, then however solid it feels, one is mainly raising empty space through air which is itself mainly empty. Within that arm there are positively and negatively charged species which interact but the overall effect is still one of a neutral charge.

As ever, discoveries raised more questions than it answered:

1. Overall, the atom is neutral but is comprised of equal numbers of charged particles but the mass of the atom could not be accounted for by the masses of the constituent protons and electrons. This meant that there had to be another, as yet undiscovered, particle of no charge and of unknown mass. This has already been discussed in Chadwick's discovery of the neutron (p.23) which is important for the path we are taking through the discoveries (as it explained the existence of the isotopes) but, chronologically, Chadwick's discovery was made after Thomson's demonstration of the existence of the electron's wave-like properties.

2. If an electron can behave as both a wave and a particle then can other particles, such as protons and atoms, behave also as waves? The answer would appear to be Yes if only because if not then why not?

But the major question still arose and had not been answered:

3. Classically, a positively charged particle orbited by a negatively charged particle should attract the negative charge and the collision would result in the emission of radiation. That clearly does not occur since, if it did then we not be here to even ask the question, so why does it not occur? Why are electrons content to exist around the nucleus, held by it, but not pulled in to it?

Erwin Schrödinger had provided the answers to questions 2 and 3.

Schrödinger's Wave Equation and the Atom's Electronic Distribution

By the mid 1920s, physicists knew that light which had been regarded as wavelike in nature also behaved as particles, which became known as *photons*, and that electrons, which had been believed to be particles, could behave as waves.

Everybody finds the concept of wave-particle duality difficult at first exposure and how a particle can behave like a wave, and vice versa, has been the subject of much discussion. It is not absolute but a general rule of thumb is that if something is smaller than the lower wavelength of visible light i.e. invisible to us then it behaves more like a wave than a particle. Conversely, if it is visible to the human eye then it behaves as a particle. When one property becomes dominant it is important to note that the other property is not lost but is just insignificant to the circumstance. It does not lose its other property but that other property becomes much less important. Mathematical descriptions of the wavelike natures of particles are known as *wave functions*, describe the state of a system at a particular time and predict the future state of the system.

The next question was:

Since waves can behave like particles and particles can behave as waves does an atom also possess a wave nature?

The answer was provided by Erwin Schrödinger who provided a wave equation to describe the atom. It is not necessary to have either the mathematical or physics knowledge to understand the consequences of the equation which are pertinent to chemistry.

Essentially, the Schrödinger equation predicts that if certain properties of a system are measured, only certain, discrete, values can be solutions to the equation. From a chemical point of view the importance of the equation is that it describes how electrons can be arranged in an atom. The mathematics is well outside of the scope of this volume, and unnecessary for chemists, but the conclusions can be summarised as follows:-

- Electrons *do* exist in regions remote from the nucleus which resonates with the Bohr model of orbits.
- There are a number of such regions which exist progressively further from the nucleus and these are now known as *energy levels*, since from the stepladder analogy, the electrons further from the nucleus **must** have more potential energy than those closer to the centre of the atom.
- Each energy level can hold a maximum number of electrons as shown below for the first four energy levels:

Energy Level	Maximum number of electrons
1	2
2	8
3	18
4	32

- When the atom is not being irradiated, it exists in its own lowest energy state, known as the *ground state*. In other words, the electrons exist in the lowest possible energy level and as the atomic number increases energy levels 1, 2, 3, 4, etc; become filled in that order. These are known as the ground states of the electrons.

- This means that the electron(s) of $_1$H and $_2$He exist in the first energy level and the electrons of atoms of atomic number 3 – 10 exist in the first two energy levels etc;

- The electrons do not circle the nucleus in one orbit, as envisaged by Rutherford nor do they *circle* the nucleus in the rings proposed by Bohr and neither do they exist in the Saturnian model proposed by Nagoako however Bohr's conception of energy levels was consistent with the Schrödinger wave equation and the solutions to the Schrödinger equation essentially detailed how the electrons are arranged within the energy levels.

- It is not possible to state exactly where the electrons exist within their energy levels but we can assign a probability to their locations. The specific regions of space which electrons can occupy are known as **orbitals**. The arrangement of the electron in the energy levels and, within the energy level, within the orbitals is known both as the *electron distribution* and as the *electronic configuration*. These two terms essentially have the same meaning.

- The orbitals are labelled using the letters **s**, **p**, **d** and **f**. There is no mystery about this as they simply relate to the appearance of different regions of absorption and emission spectra and the naming is as follows: **s**: *sharp* **p**: *principal* **d**: *diffuse* **f**: *fundamental*

Again, the *shape* of the orbitals indicates where the electron is most *likely* to be found. It is not exact and is a *probability*. Orbitals have no physical structure and they are best thought of as contour maps of the atom outlining where the electrons are most likely to be found.

We will now discuss the s – orbitals, p – orbitals and, briefly, the d – orbitals in order.

The Arrangements and Distribution of Electrons within Orbitals

S – Orbitals

- **S – orbitals** are ring-doughnut shaped. Topologically they are called toroids and, although conventionally drawn as a circle, there must be zero electron density in the centre as that is the nucleus.

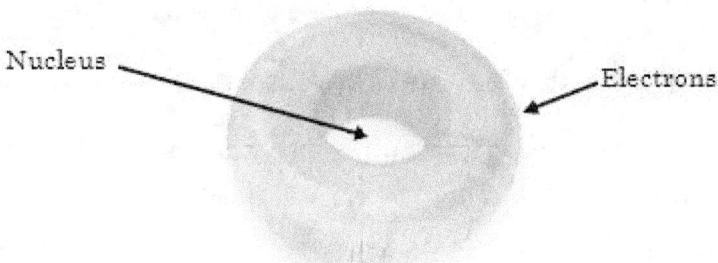

Incidentally, this is the case with all orbitals: they must have zero distribution at the centre of the atom as that is always occupied by the nucleus.

Every energy level contains an **s – orbital** and regardless of the number of energy levels it is the first orbital (region) to be filled. This means that the first two electrons fit into the 1s orbital and are described as 1s electrons and this accounts for hydrogen and helium: $_1$**H:** $1s^1$ $_2$**He:** $1s^2$

The numeral describes the energy level, the letter denotes the orbital and the superscript records the number of electrons in that orbital and hence in that energy level. Since the first energy level can only have one orbital and that orbital can only hold a maximum of two electrons, that energy level is then filled.
This is highly significant and explains the reactivity of hydrogen atoms and the complete inertness of helium and the rest of the Noble Gas atoms. It also begins to explain the reactivity of the Group I and II elements and the relative reactivity of the Group I and II elements, all of which is discussed later.

This leads us on to Period 2 of the Periodic Table and we can start with lithium.

Lithium has the atomic number 3 and so the atom has three electrons surrounding the nucleus.

- The first two electrons fit in to the 1s shell (energy level).
- The third electron must go into the second energy level (shell)

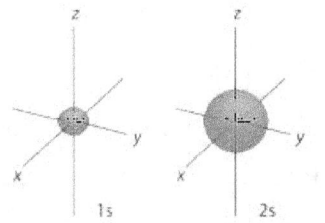

The third electron goes into the second energy level and fits into another s orbital, hence 2s, which is larger than the 1s orbital and envelopes it:

The electronic distribution of lithium ($_3$**Li**) is thus: **$1s^2$ $2s^1$**.

Since each orbital can hold up to two electrons this also leads to the conclusion that beryllium has the electron configuration: $_4$**Be: $1s^2$ $2s^2$**.

The principle that electrons fit into the orbital of lowest energy sequentially is known as the **Aufbau** (*'building up'*) **Principle** and applies to all orbitals in all energy levels. All s – orbitals are symmetrically and spherically identical but as the atom gets bigger then the s – orbitals increase in volume so for the first three periods the s – orbitals 1s, 2s, 3s can be visualised as shown on the right.

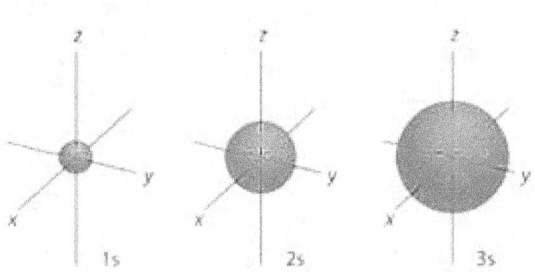

P Orbitals

As we move across the Periodic Table and encounter atoms of atomic number greater than four we can confidently state that the first four electrons have the configuration $1s^2\,2s^2$.

The question then is where do the other electrons exist?

The answer is in a beautifully symmetrical set of three orbitals, each which can hold a maximum of two electrons. They are often described as dumb-bell shaped but it is better to consider them to resemble a figure of eight.

There are three of them, orthogonal to each other as shown:

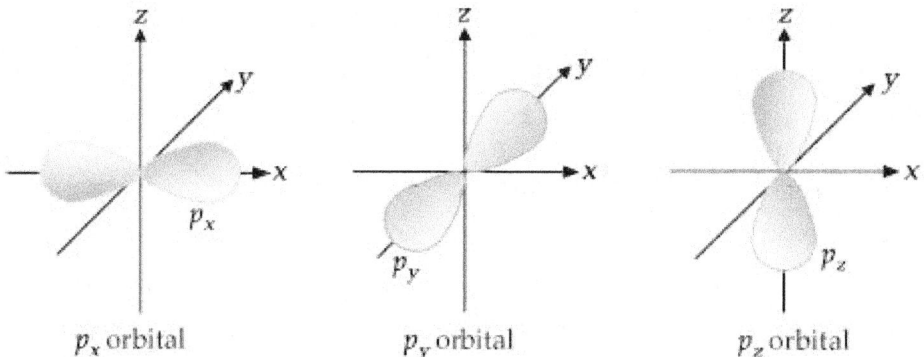

p_x orbital $\qquad\qquad$ p_y orbital $\qquad\qquad$ p_z orbital

The easiest way to envisage the x, y and z axes is to stand up, extend one hand to the front and one to the right (or left). These are the x and y – axes and your body is the z – axis. There are several important principles to absorb and remember:-

- Although the 2s and 2p orbitals exist within the same energy level, the p-orbitals (all of equal energy to each each, *mutually degenerate*) are of higher energy than the 2s orbital. This must be the case since the p-orbitals extend further from the nucleus than the 2s orbital and there is a direct correlation between distance from the nucleus and energy as shown below:

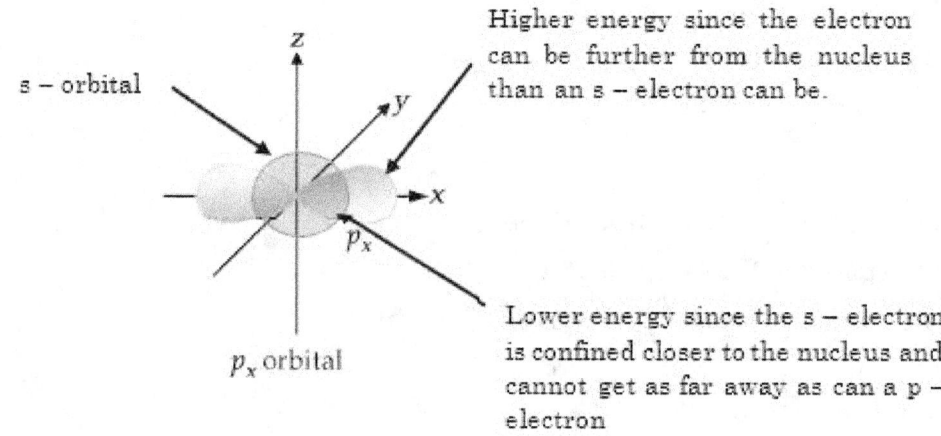

p_x orbital

s – orbital

Higher energy since the electron can be further from the nucleus than an s – electron can be.

Lower energy since the s – electron is confined closer to the nucleus and cannot get as far away as can a p – electron

This principle applies to all three 2p orbitals as shown below:

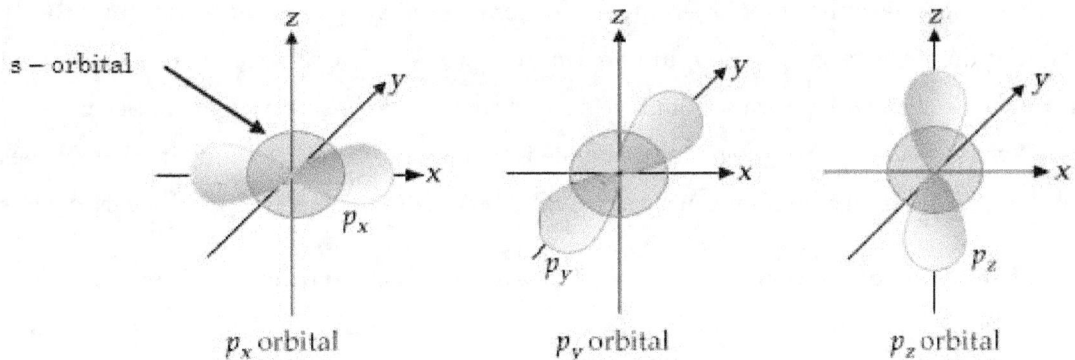

| p_x orbital | p_y orbital | p_z orbital |

Moving right from beryllium, the next six elements have their extra electrons placed in the p – orbitals.

To understand this it is essential to remember that there are three p – orbitals, which are of the same energy i.e. *mutually degenerate*, and which are denoted, arbitrarily, p_x, p_y and p_z.

Since electrons repel each other the first three electrons will exist in the three p – orbitals before they are then paired up so we can represent the electronic distribution of boron, atomic number 5, as: $1s^2\ 2s^2\ 2p_x^1$.

This indicates that the:-

- First two electrons in each energy level exist in the s – orbital;
- The fifth electron exists in one of the p – orbitals, arbitrarily denoted $2p_x$ above.

Therefore, for-;

- Carbon, $_6$C, the distribution is $1s^2\ 2s^2\ 2p_x^1\ 2p_y^1$ whilst
- Nitrogen, $_7$N, can be written as $1s^2\ 2s^2\ 2p_x^1\ 2p_y^1\ 2p_z^1$.

This is a formulation of **Hund's Rule** which states that '*the lowest-energy configuration for an atom with electrons within a set of degenerate orbitals is that having the maximum number of unpaired electrons*'.

For convenience, and provided Hund's Rule is remembered, it is usual to ignore the x, y and z – notations such that the above three notations become:

- *Boron*, $_5$B: **$1s^2\ 2s^2\ 2p^1$**
- *Carbon*, $_6$C, **$1s^2\ 2s^2\ 2p^2$**
- *Nitrogen*, $_7$N, **$1s^2\ 2s^2\ 2p^3$**

Continuing across Period 2, the electrons in the p – orbitals start pairing up and, for oxygen, fluorine and neon we have:-

- $_8$O $\quad 1s^2\ 2s^2\ 2p_x^2\ 2p_y^1\ 2p_z^1 \quad$ equivalent to $\quad 1s^2\ 2s^2\ 2p^4$
- $_9$F $\quad 1s^2\ 2s^2\ 2p_x^2\ 2p_y^2\ 2p_z^1 \quad$ equivalent to $\quad 1s^2\ 2s^2\ 2p^5$
- $_{10}$Ne $\quad 1s^2\ 2s^2\ 2p_x^2\ 2p_y^2\ 2p_z^2 \quad$ equivalent to $\quad 1s^2\ 2s^2\ 2p^6$

Nobody would deny that this is cumbersome and there are two other ways to denote the electronic distribution.

Linear notation

As a shortened version where the three p – orbitals are considered as one group so we can write the periodic electronic distribution of the second period, in two different ways. We can write it as used above but that is also tedious and it is prone to errors through a slip of the pen and so one way to record the electronic distribution is to write the symbol of the previous Noble Gas in square brackets and simply add the additional electrons afterwards. The two equivalent notations are shown below, separated by the vertical dashed line.

Element	Electronic configuration		Element	Electronic configuration	
$_3$Li	$1s^2\ 2s^1$	[He] $2s^1$	$_{11}$Na	$1s^2\ 2s^2\ 2p^6\ 3s^1$	[Ne] 3^1
$_4$Be	$1s^2\ 2s^2$	[He] $2s^2$	$_{12}$Mg	$1s^2\ 2s^2\ 2p^6\ 3s^2$	[Ne] 3^2
$_5$B	$1s^2\ 2s^2\ 2p^1$	[He] $2s^2\ 2p^1$	$_{13}$Al	$1s^2\ 2s^2\ 2p^6\ 3s^2\ 3p^1$	[Ne] $3^2\ 2p^1$
$_6$C	$1s^2\ 2s^2\ 2p^2$	[He] $2s^2\ 2p^2$	$_{14}$Si	$1s^2\ 2s^2\ 2p^6\ 3s^2\ 3p^2$	[Ne] $3^2\ 2p^2$
$_7$N	$1s^2\ 2s^2\ 2p^3$	[He] $2s^2\ 2p^3$	$_{15}$P	$1s^2\ 2s^2\ 2p^6\ 3s^2\ 3p^3$	[Ne] $3^2\ 2p^3$
$_8$O	$1s^2\ 2s^2\ 2p^4$	[He] $2s^2\ 2p^4$	$_{16}$S	$1s^2\ 2s^2\ 2p^6\ 3s^2\ 3p^4$	[Ne] $3^2\ 2p^4$
$_9$F	$1s^2\ 2s^2\ 2p^5$	[He] $2s^2\ 2p^5$	$_{17}$Cl	$1s^2\ 2s^2\ 2p^6\ 3s^2\ 3p^5$	[Ne] $3^2\ 2p^5$
$_{10}$Ne	$1s^2\ 2s^2\ 2p^6$	[He] $2s^2\ 2p^6$	$_{18}$Ar	$1s^2\ 2s^2\ 2p^6\ 3s^2\ 3p^6$	[Ne] $3^2\ 2p^6$

Visual representation (*Arrows in Boxes*)

This is way of representing electrons in an orbital which is represented by a simple box. Each electron is represented by a half headed arrow. The first electron ˙ points up whilst the second electron points down |.

Element	Atomic no.	Linear notation	Visual notation 1s	2s	2p	Electrons in shells
H	1	$1s^1$	↑			1
He	2	$1s^2$	↑↓			2
Li	3	$1s^2\ 2s^1$	↑↓	↑		2, 1
Be	4	$1s^2\ 2s^2$	↑↓	↑↓		2, 2
B	5	$1s^2\ 2s^2\ 2p^1$	↑↓	↑↓	↑	2, 3
C	6	$1s^2\ 2s^2\ 2p^2$	↑↓	↑↓	↑ ↑	2, 4
N	7	$1s^2\ 2s^2\ 2p^3$	↑↓	↑↓	↑ ↑ ↑	2, 5
O	8	$1s^2\ 2s^2\ 2p^4$	↑↓	↑↓	↑↓ ↑ ↑	2, 6
F	9	$1s^2\ 2s^2\ 2p^5$	↑↓	↑↓	↑↓ ↑↓ ↑	2, 7
Ne	10	$1s^2\ 2s^2\ 2p^6$	↑↓	↑↓	↑↓ ↑↓ ↑↓	2, 8

All of these methods of representing electronic distribution have particular value and have led to immediate explanations for well known but until then not understood, chemical phenomena:

There are many examples which demonstrate this and a few are now to be considered.

The Significance of the Noble Gases to Electronic Distributions

The *Noble Gases* were known as the *Inert Gases* until the 1960s when the first compound of xenon was formed under extreme conditions. They were, however, not discovered until the turn of the 20th century and there is a simple reason for this. Essentially, they do not do anything: they are unreactive, colourless and odourless so there was no reason to suspect that they even exist. This remained the case until Sir William Ramsay began to investigate the composition of air since the measured quantities of nitrogen, oxygen and carbon dioxide did not add up to 100% This led to the natural conclusion that air must contain at least one other gas and there was another curiosity: the density of nitrogen obtained from chemical compounds did not equal the apparent density of atmospheric nitrogen which was all that was assumed to remain after the removal of oxygen, water and carbon dioxide.

Ramsay sequentially removed oxygen , nitrogen and carbon dioxide. The oxygen was removed by passing the air over hot iron filings (forming solid iron(III) oxide), nitrogen by passing the remaining air over heated magnesium (forming solid magnesium nitride) and the carbon dioxide by bubbling the remaining air through lime water. The procedures were performed in a cycle of this chain of reactions and the air was repeatedly passed though this sequence for about three weeks. Eventually the remaining air was passed into an evacuated container and cooled under high pressure to collect the liquefied residual gases. The liquefied gas was gradually warmed up and each fraction collected separately.

The conclusions were stunning.

The gas with the lowest boiling point vaporised first and was collected. When analysed spectroscopically it was found to emit lines previously only known from spectroscopic analysis of sunlight. Ramsay had isolated the element previously only known in sunlight and named it *helium* after the Greek name for the sun: Helios. Ramsay isolated four other fractional gases and he named one a*rgon*, a combination of two Greek words for *lazy gas*. The others he discovered were neon, krypton and xenon. To Mendeleev's dismay, Ramsay then proposed a new column of the Periodic Table and he predicted the existence of *radon* which was subsequently discovered as a by-product of the radioactive decay of radium.

For the purposes of this volume, the significance of the newly discovered group was identified by Gilbert 'G.N.' Lewis. In 1916, Lewis, using the Bohr model observed that the inert gases all had two features in common: **all are unreactive and all possess a filled set of energy levels.**

Lewis suggested that there must be a correlation between lack of reactivity and the filled energy levels and proposed that filled energy levels conferred a particular stability on the atom hence leading to a lack of reactivity.

The solutions to the Schrödinger equation, which for our purposes are the electronic configurations, enable us to explain why Group I (alkali metals) elements only ever form singly positive ions e.g. for sodium Na^+ and never an Na^{2+} ion or an ion of even higher charge.

Following on from this we can also now consider why the halogens only form single, negatively charged, ions.

Considering the alkali metals first;

The chemical formation of sodium chloride is a magnificent demonstration of the power of chemistry:

A grey metal that will burn us (sodium) reacts with a yellow – green gas which can kill us (chlorine) to produce a white solid which we can safely put, within reasonable amounts on our food, sodium chloride, and it turned out that this was due simply to transfer of a single electron from each sodium atom to each chlorine atom of the Cl_2 molecule.

The word equation and the balanced symbolic equations are as follows

$$sodium \quad + \quad chlorine \quad \longrightarrow \quad sodium\ chloride$$

$$2\ Na\ (s) \quad + \quad Cl_2\ (g) \quad \longrightarrow \quad 2NaCl(s)$$

With the determination of the electronic distribution within the orbitals which were solutions to the Schrödinger wave equation, the reason for both suddenly became clear.

Alkali metals

The *sodium* atom, atomic number 11, has the electronic distribution $1s^2\ 2s^2\ 2p^6\ 3s^1$

1s	2s	2p			3s
↑↓	↑↓	↑↓	↑↓	↑↓	↑

If it loses one electron then it assumes the electronic distribution of the *neon* atom:

1s	2s	2p		
↑↓	↑↓	↑↓	↑↓	↑↓

The sodium ion becomes unreactive once the electron shells are all filled; in this case electron shells 1 and 2.

The concept of stability conferred by a Noble Gas electronic configuration was called by Lewis, the *Octet Rule* since the second energy level contains eight electrons explains the reactivity of the sodium atom, its stability as a Na^+ ion and its then complete lack of reactivity.

This explanation is reinforced by considering the same reaction with potassium which as the electronic configuration $1s^2\ 2s^2\ 2p^6\ 3s^2\ 3p^6\ 4s^1$ and as shown visually below.

1s	2s	2p			3s	3p			4s
↑↓	↑↓	↑↓	↑↓	↑↓	↑↓	↑↓	↑↓	↑↓	↑

One electron exists in an outer shell and when that is lost the K^+ ion has the same electronic configuration as argon and becomes unreactive. It is important to note that, although the 3rd energy level also contains five d – orbitals these are of higher energy than the 4s orbital which is filled before the d – orbitals are used.

The same occurs with lithium which loses one electron to become Li^+ which has the electronic configuration of helium.

It is now clear then that reactivity depends on the number of outer electrons.

When the atom or ion contains a filled set of electron shells then it goes quiet and becomes unreactive. There is a completely analogous set of descriptions of the behaviour of the Group II atoms which form 2+ ions.

Halogens

Moving to the other side of the Periodic Table we can now consider the **halogens**.

For this purpose, the three most important halogens are fluorine, chlorine and bromine. All three atoms form single negatively charged ions **X⁻** where **X** represents the atom.

The electronic configuration of the atoms are as follows:-

Element	Atomic number	Electronic configuration
Fluorine	9	$1s^2\ 2s^2\ 2p^5$
Chlorine	17	$1s^2\ 2s^2\ 2p^6\ 3s^2\ 3p^5$
Bromine	35	$1s^2\ 2s^2\ 2p^6\ 3s^2\ 3p^6\ 4s^2\ 4p^6\ 5s^2\ 5p^5$

The halogens form only single charged anions, X⁻, adopting the configuration of the subsequent Noble Gas.

Fluorine adopts the electronic configuration of *neon*, **chlorine** that of *argon* and **bromine** that of *krypton*:

Atom	Ion	Noble Gas equivalent
Fluorine (F) $1s^2\ 2s^2\ 2p^5$	**Fluoride (F⁻)** $1s^2\ 2s^2\ 2p^6$	Neon
Chlorine (Cl) $1s^2\ 2s^2\ 2p^6\ 3s^2\ 3p^5$	**Chloride (Cl⁻)** $1s^2\ 2s^2\ 2p^6\ 3s^2\ 3p^6$	Argon
Bromine (Br) $1s^2\ 2s^2\ 2p^6\ 3s^2\ 3p^6\ 4s^2\ 4p^6\ 5s^2\ 5p^5$	**Bromide (Br⁻)** $1s^2\ 2s^2\ 2p^6\ 3s^2\ 3p^6\ 4s^2\ 4p^6\ 5s^2\ 5p^6$	Krypton

This explains the instability of sodium (and the rest of the Group I, alkali metals) to create a stable and chemical unreactive material, *sodium chloride*, which we use in cooking as common, or table, salt.

It also explained why chlorine does not exist nature as single atoms but as a diatomic molecule, Cl_2, which will be addressed later.

These analyses can be extended to Group II, the alkaline earth metals, which form **M^{2+}** ions where **M** is the element and and to Group VI with the best example being oxygen.

Oxygen has the atomic number 8 and so has the electronic configuration, $1s^2\ 2s^2\ 2p^4$ also visually depicted as:-

1s	2s	2p		
↑↓	↑↓	↑↓	↑	↑

To adopt the configuration of the next Noble Gas, **neon**, the oxygen atom needs to acquire two electrons.

It does not matter where these electrons come from and it can be exemplified by two reactions:

1.   If sodium is heated in oxygen then the product is sodium oxide Na_2O.

sodium	+	**oxygen**	$\longrightarrow$	**sodium oxide**
light grey metal		colourless gas		white/grey solid

$$2\,Na \quad + \quad O_2 \quad \longrightarrow \quad Na_2O$$

Two sodium atoms each want to lose an electron and oxygen wants to gain two.

Due to this transfer of electrons both sodium and oxygen adopt the stable, and consequently unreactive electronic distribution of neon.   This explains the reaction and the formula of the product.

2.   The second example is that of magnesium combusting in oxygen.

magnesium	+	**oxygen**	$\longrightarrow$	**magnesium oxide**
light grey metal		colourless gas		white/grey solid

$$2\,Mg \quad + \quad O_2 \quad \longrightarrow \quad 2\,MgO$$

Each magnesium atom has two outer electrons which it wants to lose to assume the electronic distribution of neon whilst oxygen wants to acquire them to also adopt that stable configuration.

The same rules explain the formation of all ions of Group I (alkali metals) and Group II (alkaline earth) metals and why the former acquire a single positive charge whilst the latter form stable 2^+ ions.   There is a consistency in all of these reactions and the theory demonstrates why metals lose electrons to acquire the electronic distribution of the previous noble gas configuration (as arranged in the Periodic Table) whilst non – metals gain electrons to acquire the distribution of electrons of the Noble Gas following the non-metal.

Bonding involving the transfer of one or more electrons from one atom to another results in compounds of high melting and boiling points and are water soluble.   This type of bonding is ***ionic bonding*** and, although not particularly relevant to our investigation as we are looking at organic compounds, it does explain two of three phenomena observed with chemical substances:

1.   Reaction between a metal and a non-metal produces a compound with high melting point, dissolves in water and when molten conducts electricity.   This is also due to the fact that the product of the reaction comprises positive and negatively charged ions which attract each other and hold themselves together very tightly but when dissolved in water can carry an electric current.   These simple ionic compounds form between a metal on the left of the Periodic Table and a non-metal on the right hands side.

2.   No compounds of two metals exist (alloys of metals do, course, exist but they are physically mixed and not chemically combined and such compounds cannot exist since, if both metals wish to lsoe their outer electrons, where are they to go?

3.   The third, so far unexplained phenomenon is that of low boiling, water insoluble organic compounds, formed from elements in the middle of the Periodic Table and, of course, hydrogen, which cannot be explained by the above electron transfer so there must be some other process at work and this is investigated next.

Chapter V

Bonding concepts and principles

This chapter considers the vast majority of chemical compounds which can exist as solids, liquids or gases, tend to have no or limited water solubility and, unlike ionic compounds, do not conduct electricity but do dissolve in neutral solvents.

In essence, they are the antithesis of ionic compounds, and contain elements present in the Middle of the Periodic Table as well, of course, hydrogen. The properties of these compounds and the way that the atoms bond together can be explained by referring back again to the electronic configuration of these elements.

This chapter discusses the central region of the Periodic Table and introduces, arithmetically simple, but extremely profound concepts.

We consider and investigate the Central Main Groups of the Periodic Table and develop the concepts of:-

- Covalent Bonding
- Effective Atomic Number
- Electronegativity
- Bond polarities
- Bond Strength

all of which derive from the theories which have been discussed in the earlier chapters.

The Central Groups of the Periodic Table

Questions About Bonding

The examples given in the previous chapter all involve electron transfer from a metal to a non – metal and explain the phenomenon of *ionic bonding*. Ionically bonded compounds have a number of identifying features which we can now see arise from electron transfer. They dissolve in water, conduct electricity when molten and have high melting points, the last demonstrating that the ions are held tightly together.

The following questions then naturally arise:

- Why do innumerable compounds and molecules have low melting and boiling points?
- Why do many compounds not conduct electricity?
- Why do many not dissolve in water but do dissolve in organic solvents?
- What process is involved when compounds and molecules containing two identical atoms bond to each other e.g. hydrogen (H_2), nitrogen (N_2), chlorine (Cl_2) and the carbon atoms in ethane (C_2H_4) etc; ?

Taking the last point first, it was well known that different atoms would form different numbers of bonds. For example, hydrogen only ever makes one bond, nitrogen always makes three and carbon always makes four. This tendency became known as the element's *valency* which was conceived by Edward Frankland and gave rise to the concept of multiple bonds which was the only way to make sense of the bonding in organic molecules.

The following sections discuss the sharing of electrons and considers its occurrence in the hydrogen, oxygen, nitrogen, ammonia and chlorine molecules before moving to consider the bonding in the compounds of most importance to this series which are organic compounds, those containing carbon and hydrogen atoms at the least.

Covalent bonding

The Hydrogen Atom

Hydrogen has one electron in its first shell hence its configuration is $1s^1$.

It, therefore, has three choices since it can:-

1. Lose its electron to form the hydrogen ion (H^+),
2. Gain an electron to form a hydride ion (H^-), or
3. Share its electron.

Considering these possibilities in order,

1. When a hydrogen atom loses its electron, it becomes what chemists refer to as a hydrogen ion (H^+) but which from a physicist's point of view is simply a proton. Due to the enormous ratio of charge to mass, a proton cannot exist on its own. In water it bonds with a water molecule to form a hydronium ion, H_3O^+ but since this series is concerned with organic compounds the fascinating aqueous chemistry of the hydronium ion is beyond our scope here.
2. A hydrogen atom can form a hydride ion and some hydride compounds are known. There are relatively few of them and are not relevant here.
3. A hydrogen atom can share its electron with another atom which is content to also share an electron. This is known as *covalent bonding*.

The clearest example that situation #3 is the most common is demonstrated by the millions of organic compounds which are gaseous, liquid or solids with low melting points.

Taking hydrogen, nitrogen, fluorine and carbon in this order:

Bonding in the hydrogen (H₂) molecule

Hydrogen, atomic number one, has only one electron in its 1s orbital which can hold two. If the two hydrogen atoms share their respective electron with the other other one then each atom achieves two electrons and acquires the stable helium structure:

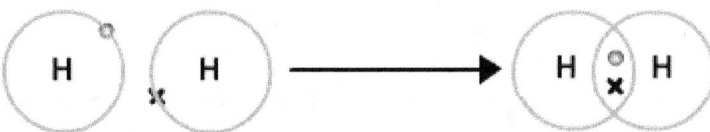

Since there are now no vacant positions in the first energy level we can now understand why hydrogen forms only one bond and why the element exists naturally as the H_2 molecule.

Sharing electrons is known as *covalent bonding*.

Bonding in the ammonia (NH₃) molecule

Having understood the concept of sharing electrons we can now explain why ammonia has the chemical formula NH₃.

Atomic nitrogen ($_7$N) has the electronic configuration $1s^2\ 2s^2\ 2p^3$ which is most clearly represented by the *arrows in boxes* method.

$$\begin{array}{ccc} \textbf{1s} & \textbf{2s} & \textbf{2p} \\ \boxed{\uparrow\downarrow} & \boxed{\uparrow\downarrow} & \boxed{\uparrow}\ \boxed{\uparrow}\ \boxed{\uparrow} \end{array}$$

If we assume that the *Octet Rule* applies in all cases then, to achieve a Noble Gas configuration, nitrogen must either lose five electrons or gain three.

Losing five electrons is clearly unfeasible since they have to go somewhere but gaining three electrons through sharing electrons with hydrogen atoms appears reasonable and explains the molecular formula of ammonia (NH₃).

The nitrogen atom shares three of its electrons, those in the 2p orbital, with the three hydrogen atoms (one electron with each hydrogen atom) meaning that the hydrogen atom now has a filled energy level with the electronic configuration of helium.

If each hydrogen atom reciprocates and shares its electron with nitrogen then nitrogen too secures a Noble Gas configuration (that of neon) and the bonding in ammonia (NH₃) can be drawn as shown below.

The bonding in ammonia (NH₃):

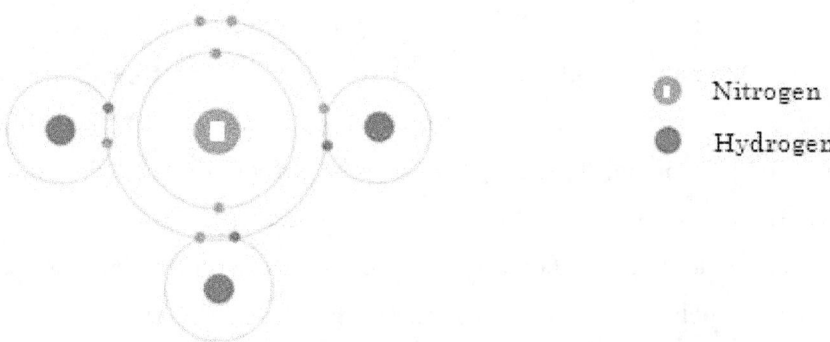

○ Nitrogen

● Hydrogen

To reiterate, in the case of each atom, **nitrogen** achieves the configuration of *neon* whilst each **hydrogen** atom achieves that of *helium*.

Bonding in the fluorine (F₂) molecule

The fluorine ($_9$F) atom has the electronic configuration which can be written as $1s^2\ 2s^2\ 2p^5$ or most clearly, for our purposes, using arrows in boxes:

1s   2s   2p

⇅	⇅	⇅	⇅	↑

It is clear that to fill the 2p orbital, a fluorine atom needs one electron since it is clearly unfeasible to expect the atom to lose seven. Clearly, in the F₂ molecule each fluorine atom shares one electron with its counterpart and this explains its molecular formula.

Bonding of the carbon atom

We are now in a position to understanding the bonding of a ***carbon*** atom which, with the atomic number six, has the following electronic distribution:

1s   2s   2p

⇅	⇅	↑	↑	

To achieve a stable electronic configuration of only filled electron energy levels there are three options: either carbon loses four electrons or it gains four by sharing or by stealing. In principle, all are possible but if the carbon atom lost four electrons it would form a C^{4+} ion and assume the electronic configuration of helium. Such compounds would then be extremely strong with high melting and boiling points. In fact, no such compounds are known and as many carbon containing compounds are liquids or gaseous since they have low melting and boiling points, the bonding in these compounds cannot be ionic and must be covalent i.e. the valence electrons are shared.

The first atom to share an electron with carbon will deposit that electron in the empty 2p orbital. This is because electrons repel each other and try to keep as far apart as possible and it is only after then that the electrons start pairing up. This fully accounts for carbon forming four bonds since it needs four electrons and explains carbon – carbon single, double and triple bonds. It does not however, explain why carbon in methane forms four equivalent bonds of equal length and hence of equal strength. That is due to a phenomenon called *hybridisation* which is largely beyond the scope of the main text of this book. Essentially, however, the overlapping 2s and 2p orbitals shake themselves up a little to create four equivalent orbitals. These orbitals are known as sp³ hybrid orbitals simply to indicate that they are created from *one* **s** orbital and *three* **p** orbitals. It is well worth studying this fascinating phenomenon but further consideration is unnecessary for our purposes.

Taking stock

In a mere few dozen pages, we have come a long way in this volume from the concept of a hard, impenetrable atom to a fundamental understanding of the dense nucleus in the centre of an atom which is mainly empty and has the electrons surrounding it in specific regions.

The existence of three sub-atomic particles have been discussed and we now understand why some compounds have very high melting and boiling points whilst others are liquids or gases and room temperature and do not dissolve in water. The arrangement of the Periodic Table has been validated, the reason why Groups of elements behave in a similar manner is clear and the formation of ions of particular charges but no other is clear. Whilst the mathematics involved in solving the Schrodinger equation are beyond the scope of this volume the solutions to the equation are profound and can be summarised simply as a set of rules. So far we know that:-

- The positive charges exist in a small dense area in the centre of the atom, the nucleus.
- The extra mass, comprised of neutrons, exists in the same region of the atom.
- Most of the volume is empty space, known as *free space*.
- Electrons exist in energy levels i.e. regions of space at different distances from the nucleus.
- They do not circle the nucleus like a train of electrons, and within each energy level they can only exist within specific regions.
- The first energy level only contains the spherical s – orbital. This shaped orbital is always the first filled in each energy level.
- The second energy level can hold a maximum of eight electrons. The first orbital to filled is an s orbital and there are then three p – orbitals.
- All these regions overlap e.g. the 3s orbital is greater in volume and encompasses the 2s – orbital which similarly is larger than the 1s – orbital.
- Later energy levels are comprised of a single s – orbital, three p – orbitals, five d – orbitals etc;

*When examining the structure of metal atoms and ions, it is important to remember that orbitals are not physical, tangible objects. They are merely a description of the **probability** of finding an electron in a particular region of the atom. Although we talk about shape, volume and size, orbitals are best thought of as a map showing the likelihood of the finding an electron there.*

- The three p – orbitals in energy level 2 are all identical in size in volume and the only difference between them is their orientation in space, along the x, y and z – axes.
- The p – orbitals overlap in space with the s – orbital but, as discussed previously, as their lobes (the regions) extend further than the 2s – orbital, they must be of higher energy than the 2s orbital.
- The three p – orbitals in energy level 3 are all identical in size in volume and the only difference between them is their orientation in space, along the x, y and z – axes. They differ from the 2p orbitals in size only. Existing in a higher energy level i.e. further from the nucleus they are merely larger and of higher energy than the 2p analogues.

The chemistry of the first twenty elements can be explained by the electronic configuration as shown below:

Element	Linear notation	1s	2s	2p			3s	3p			4s
$_1$H	$1s^1$	↑									
$_2$He	$1s^2$	↑↓									
$_3$Li	$1s^2\,2s^1$	↑↓	↑								
$_4$Be	$1s^2\,2s^2$	↑↓	↑↓								
$_5$B	$1s^2\,2s^2\,2p^1$	↑↓	↑↓	↑							
$_6$C	$1s^2\,2s^2\,2p^2$	↑↓	↑↓	↑	↑						
$_7$N	$1s^2\,2s^2\,2p^3$	↑↓	↑↓	↑	↑	↑					
$_8$O	$1s^2\,2s^2\,2p^4$	↑↓	↑↓	↑↓	↑	↑					
$_9$F	$1s^2\,2s^2\,2p^5$	↑↓	↑↓	↑↓	↑↓	↑					
$_{10}$Ne	$1s^2\,2s^2\,2p^6$	↑↓	↑↓	↑↓	↑↓	↑↓					
$_{11}$Na	$1s^2\,2s^2\,2p^6\,3s^1$	↑↓	↑↓	↑↓	↑↓	↑↓	↑				
$_{12}$Mg	$1s^2\,2s^2\,2p^6\,3s^2$	↑↓	↑↓	↑↓	↑↓	↑↓	↑↓				
$_{13}$Al	$1s^2\,2s^2\,2p^6\,3s^2\,3p^1$	↑↓	↑↓	↑↓	↑↓	↑↓	↑↓	↑			
$_{14}$Si	$1s^2\,2s^2\,2p^6\,3s^2\,3p^2$	↑↓	↑↓	↑↓	↑↓	↑↓	↑↓	↑	↑		
$_{15}$P	$1s^2\,2s^2\,2p^6\,3s^2\,3p^3$	↑↓	↑↓	↑↓	↑↓	↑↓	↑↓	↑	↑	↑	
$_{16}$S	$1s^2\,2s^2\,2p^6\,3s^2\,3p^4$	↑↓	↑↓	↑↓	↑↓	↑↓	↑↓	↑↓	↑	↑	
$_{17}$Cl	$1s^2\,2s^2\,2p^6\,3s^2\,3p^5$	↑↓	↑↓	↑↓	↑↓	↑↓	↑↓	↑↓	↑↓	↑	
$_{18}$Ar	$1s^2\,2s^2\,2p^6\,3s^2\,3p^6$	↑↓	↑↓	↑↓	↑↓	↑↓	↑↓	↑↓	↑↓	↑↓	
$_{19}$K	$1s^2\,2s^2\,2p^6\,3s^2\,3p^6\,4s^1$	↑↓	↑↓	↑↓	↑↓	↑↓	↑↓	↑↓	↑↓	↑↓	↑
$_{20}$Ca	$1s^2\,2s^2\,2p^6\,3s^2\,3p^6\,4s^2$	↑↓	↑↓	↑↓	↑↓	↑↓	↑↓	↑↓	↑↓	↑↓	↑↓

It is tempting to consider the d – orbitals and the f – orbitals in the transition elements. Since however the purpose of this volume is to discuss the atomic structure in a sufficient depth to begin to analyse simple organic molecules (Volume II) and natural flavourings (Volume III) the descriptions of transition elements and their bonding, although fascinating, are not relevant for the spectroscopic and spectrometric analysis of organic molecules. The elements present in the molecules analysed in Volumes II and III all contain carbon and hydrogen whilst some contain, variously, oxygen, nitrogen and sulfur but there are a number of concepts, pertinent to these three volumes, which must be considered next. These are *effective atomic number*, *electronegativity* and *polarity* which are discussed next.

Effective Atomic Number, Electronegativity and Polarity

As a reminder: the *valence electrons* are only those in the outermost shell (energy level) of the atom. All other electrons in the inner energy levels are referred to as *core electrons*. The core electrons, although their presence is essential for the structure of the atom, play absolutely no part in bonding and can be considered as one group and essentially ignored.

To recap yet again,

- The *atomic number* of an atom or ion is the number of protons in the nucleus.
- In an atom which is **neutral**, the number of electrons surrounding the atom is equal to the atomic number (the number of positive charges in the nucleus).
- In a negatively charged ion the number of electrons equals the atomic number *plus* the number of electrons responsible for the negative charge and the opposite is true for the positively charged ions.

One pertinent and relevant question, however, is why do some atoms steal valence electrons (N, O, Cl etc;) whilst others, the alkali and alkaline earth metals – Groups I and II particularly, effectively throw their valence electrons away? Moreover, the situation appears complicated by atoms such as carbon which neither steal nor throw away their valence electrons but simply opt to share them. There has to be a reason for this and the explanation could be phenomenally complicated were it not for the concept of the *effective atomic number*, a deceptively simply but extremely powerful concept which we will consider next.

Effective Atomic Number

If we consider the valence electrons as one group, whether there is only one or more, *and* the core electrons as another group then we can perform a simple arithmetic calculation which becomes extremely revealing.

The assumption that we make is that every inner electron cancels out a proton in the nucleus.

This is not strictly true since the electrons also repel each other but are held in place by the, positively charged, nucleus but works beautifully *and* is valid to a large extent since the inner electrons *do* shield the valence electrons from the nucleus.

The calculation is straightforward and can be done in the head:

1. Write down the atomic number (A_R);
2. Sum the number of electrons in the **filled** inner shells (the core electrons);
3. Subtract the number of core electrons from the atomic number i.e.

   the answer from step 1 – the answer from step 2

4. The resulting integer is the effective atomic number.

Taking the alkali metals first.

Group I

If we consider the *lithium* atom ($_3$Li i.e. *$1s^2\ 2s^1$*).

The lithium atom has ***three*** protons in the nucleus and only one filled energy level, or shell, containing ***two*** electrons. Therefore, the calculation is

Effective atomic number = Atomic number – core electrons = 3 – 2 = +1

since there are TWO electrons in the inner shell and one in the valence shell, meaning that the *lithium* atom has an *effective atomic number* of +1.

Likewise, if we consider the *sodium* atom ($_{11}$Na, *$1s^2\ 2s^2\ 2p^6\ 3s^1$*) the calculation becomes:

Effective atomic number = Atomic number – core electrons = 11 – 10 = +1

since the atom has 10 electrons in its first two shells and one electron in the third shell. The same calculation works with the rest of the Group I elements (alkali metals) and all have an ***effective atomic number*** of +1.

The $_1$H hydrogen atom

We can, however, also consider the $_1$H atom. This atom has one proton in the nucleus and, despite having one electron it is in an incomplete shell and so the atom has ZERO filled shells. This makes the calculation:

Effective atomic number = Atomic number – core electrons = 1 – 0 = +1

which explains why some versions of the Periodic Table place the hydrogen element at the top of, or slightly above, Group I.

Group II

Moving on to the alkaline earth elements (Group II) and only considering the first three:

Using the usual formula, the calculations are as follows:-

Element	Effective atomic number
$_4$**Be** (*$1s^2$* **$2s^2$**)	Atomic number – core electrons = 4 – 2 = **+2**
$_{12}$**Mg** (*$1s^2\ 2s^2\ 2p^6$* **$3s^2$**)	Atomic number – core electrons = 12 – 10 = **+2**
$_{20}$**Ca** (*$1s^2\ 2s^2\ 2p^6\ 3s^2\ 3p^6$* **$4s^2$**)	Atomic number – core electrons = 20 – 18 = **+2**

In the above electronic configuration, the core electrons are written in *italics* whilst the valence electrons are highlighted in **bold**. This means that the alkaline earth metals have valence electrons which are held by the nucleus by twice the force of that exerted by the nuclei of the atoms of Group I. This shows that the method works since Group II elements react in a similar manner to Group I analogues *only more slowly*. The *reduced reactivity* is due to the Group II atom finding it harder to throw away the outer electrons than Group I elements since the valence electrons are more tightly held. The calculation works for all remaining elements of that Group and we can now consider the calculations for the right hand side Group VI and VII elements.

Group VII

Jumping across to Group VII (halogens) and considering only the first three: $_9F$, $_{17}Cl$, $_{35}Br$

- $_9F$ **Effective atomic number** = Atomic number − core electrons = $9 - 2 = +7$
- $_{17}Cl$ **Effective atomic number** = Atomic number − core electrons = $17 - 15 = +7$
- $_{35}Br$ **Effective atomic number** = Atomic number − core electrons = $35 - 33 = +7$

Again, all elements in Group VII are found to have an effective atomic number of +7. This is incredibly significant since it means that the valence electrons are held **SEVEN TIMES** more tightly than those of Group I elements.

Group VI

We can perform the same calculations again with oxygen and sulfur etc; Taking only these two into account (although the calculation is the same for the rest of the Group, those elements are not present in the molecules we will consider in Volumes II and III).

- $_8O$ **Effective atomic number** = Atomic number − core electrons = $8 - 2 = +6$
- $_{16}S$ **Effective atomic number** = Atomic number − core electrons = $16 - 10 = +6$

This means that oxygen and sulfur have **SIX TIMES** the ability of a group I element to attract an electron or hold on to it but are less of an electron thief than their analogous atoms in Group VII and explains the slightly lower rate of reactivity in similar reactions.

Two other Groups of high significance are the Noble Gases and Group IV.

The Noble Gases

Considering only $_2He$, $_{10}Ne$ and $_{18}Ar$

- $_2He$ Helium has two protons in its nucleus and only two electrons in its, sole, energy level.
  This means that the calculation becomes
  Effective atomic number = Atomic number − core electrons = $2 - 2 = 0$

For neon, argon and krypton, the calculations are as follows:

- $_{10}Ne$ **Effective atomic number** = Atomic number − core electrons = $10 - 10 = 0$
- $_{18}Ar$ **Effective atomic number** = Atomic number − core electrons = $18 - 18 = 0$
- $_{36}Kr$ **Effective atomic number** = Atomic number − core electrons = $36 - 36 = 0$

The formula works for the rest of the the Noble Gases and the *effective atomic number* of **ZERO** demonstrates that the atoms have no ability whatsoever to steal i.e. attract electrons and hence explains their entire lack of reactivity.

Group IV

For our purposes though, other than hydrogen, the most important group is **Group IV** which includes the element essential for life, carbon. Considering only carbon as that is the only Group IV element present in the molecules considered in Volumes II and III, carbon has the electronic distribution $1s^2\ 2s^2\ 2p^2$ i.e. it has two inner (core) electrons ($1s^2$) and four valence electrons ($2s^2\ 2p^2$) meaning that the calculation becomes

$$\textbf{Effective atomic number} = \text{Atomic number} - \text{core electrons} = 6 - 2 = +4$$

On the scale of $0-7$, the value of 4 means that the atom has no more tendency to throw away electrons than it does to steal them. This is why carbon forms covalently bonded compounds as it is quite content to share electrons.

Summary

In summary, Group I elements have an effective atomic number of +1, Group II elements have an effective atomic number of double that, Group VI elements have a value of +6 and so on and take stock of our discoveries.

It is now possible to tabulate the effective atomic numbers of the main group elements within the Periodic Table and it becomes clear that the effective atomic number equals the Group number with the exception of the Noble Gases, unless of course, Group VIII is labelled Group 0 (both numbers being used in the table below).

Group:	I	II	III	IV	V	VI	VII	VIII (0)
Effective atomic number:	+1	+2	+3	+4	+5	+6	+7	0
Period 1	H							He
Period 2	Li	Be	B	C	N	O	F	Ne
Period 3	Na	Mg	Al	Si	P	S	Cl	Ar
Period 4	K	Ca	Ga	Ge	As	Se	Br	Kr

Since the Noble Gases, traditionally placed on the right hand side of the Periodic Table, have an effective atomic number of 0 this does validate those chemists who supported placing the group on the left hand side but that is a minor point compared to the benefits of the calculation. Frankly however it appears a moot point and not worth wasting time debating and so most chemists place it on the right side, essentially just out of the way.

We can now consider the implications of the effective atomic number which explains much of chemical bonding, both ionic and covalent bonding, using the concepts of *electronegativity* and *polarity*.

Consequences of the Effective Atomic Number

We can now understand why the reaction of sodium and chlorine occurs. The reaction is written, of course as shown before, e.g.:-

$$\text{sodium} \quad + \quad \text{chlorine} \quad \longrightarrow \quad \text{sodium chloride}$$
$$2\,Na\,(s) \quad + \quad Cl_2\,(g) \quad \longrightarrow \quad 2\,NaCl\,(s)$$

whilst the equivalent reaction of magnesium and chlorine which also occurs but more slowly is:-

$$\text{magnesium} \quad + \quad \text{chlorine} \quad \longrightarrow \quad \text{magnesium chloride}$$
$$Mg\,(s) \quad + \quad Cl_2\,(g) \quad \longrightarrow \quad MgCl_2\,(s)$$

The explanation is as follows:

Sodium and chlorine

* Sodium has an effective atomic number of +1. It has only one outer electron to lose to enable it to adopt the electronic configuration of neon.

* Chlorine has an effective atomic number of +7. It only needs one electron to fill its outer shell and adopt the electronic configuration of argon.

* Sodium has **ONE** electron to lose and chlorine only needs one. Since chlorine has **SEVEN** times the power of sodium to grab the electron the reaction happens quickly and readily, forming the compound of formula NaCl.

* The resulting ions, Na^+ and Cl^- attract each other forming a strong crystalline lattice with a high melting point and it is noteworthy that, when molten it conducts electricity and it is readily soluble in liquid water, a non – ionic substance.

Magnesium and chlorine

* Magnesium has two outer electrons it can lose to adopt the stable electronic configuration of neon.

* Since magnesium has two electrons to lose and since chlorine only wants one electron, the reaction requires two chlorine atoms for each magnesium atom – each chlorine receiving one of the two electrons that magnesium wants to lose and so the formula of the resultant product is $MgCl_2$.

* The Mg – Cl bond strength is greater than that of Na – Cl which is explained by the fact magnesium ion has twice the positive charge of sodium.

* Magnesium chloride also has a high melting point but one which is slightly lower than sodium chloride since the crystalline lattice is not as tightly packed as sodium chloride due to the presence of two chloride ions for each magnesium ion.

* Magnesium chloride is also soluble in water.

The next question is

Why do ionic compounds such as sodium chloride and magnesium chloride dissolve in a neutral liquid such as water?

The answer lies in the *effective atomic number* and *electronegativity* which cause ***bond polarity***. We now understand effective atomic number and so both electronegativity and bond polarity are discussed next.

Electronegativity

Electronegativity is defined as *the ability of an atom to attract a pair of bonding electrons towards itself.*

It explains why some elements steal electrons, others let them go and some share them. It also explains the ability of water, which even though covalently bonded, is able to dissolve ionic compounds.

This is one instance where there is no opposite definition since *electropositivity* would be the ability of an atom to repel a pair of bonding electrons. Such a definition is meaningless as no atom does that but remember that negatively ions do attempt to do that. The closest an atom comes to repelling electrons is to show no interest e.g. the Noble Gases.

A number of chemists attempted to create a scale of electronegativities but the one which is most useful was created by the great physical chemist, Linus Pauling, double Nobel Laureate:- in Chemistry (1954) and for Peace (1962). Pauling scaled all electronegativities on a scale of 0 – 4 as shown below:

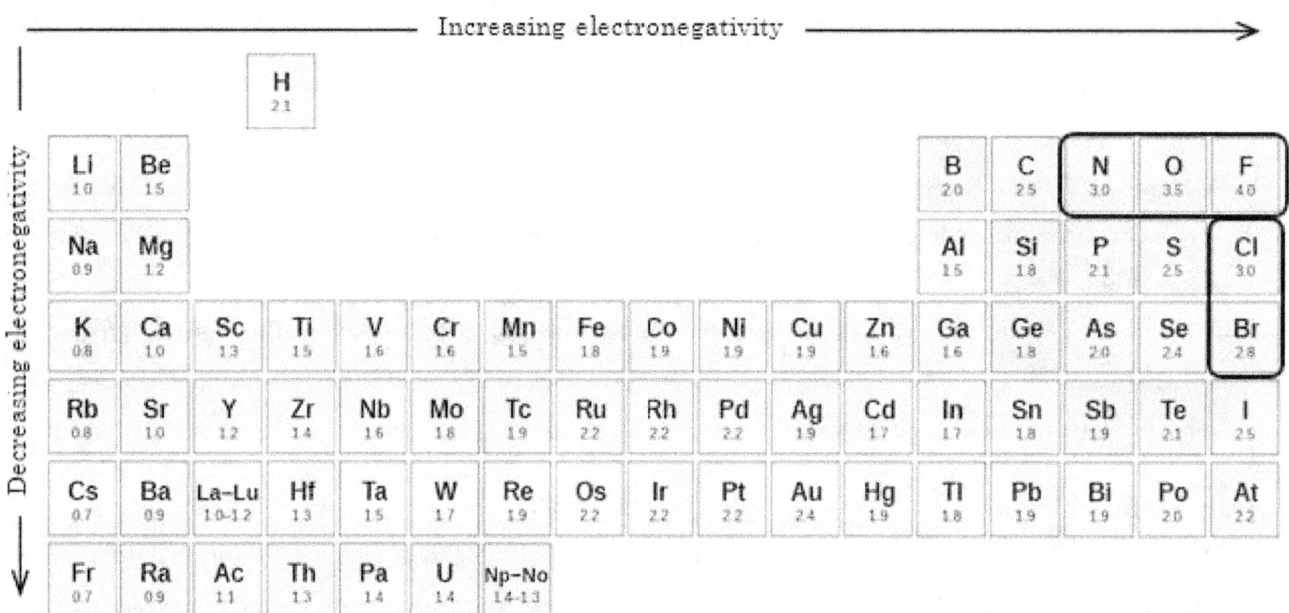

There are five elements with significant electronegativity: **nitrogen, oxygen, fluorine, chlorine and bromine** as highlighted above and the range was set such that the electronegativity of of hydrogen is 2.1. When asked why he did that and why he used a scale of 0 – 4, Pauling's response was *"I can't remember!"*

Pauling calculated his electronegativity scale on the basis of bond energies i.e. the energy required to break the bond but we can visualise simply as a function of the effective atomic number and the size of the atom:

- The higher the effective atomic number the greater the ability of that atom's nucleus to pull the bonding pair of electrons towards itself however;
- This power is affected by the number of energy levels of that atom. The larger the atom, the greater the number of core, inner, electrons that can shield the outside world from the nucleus and so;
- This means that the most electronegative elements are those with the **highest** *effective atomic number* and as **small** an atomic size i.e. as few inner energy levels as possible.

It is no coincidence that the five most electronegative elements are all in the top right hand region of the Periodic Table.

Putting this together, the element with the highest effective atomic number and the smallest size is *fluorine* and so fluorine is the most electronegative element. The other elements with a high effective atomic number and a relatively small size are the ones with high effective atomic number and a small size i.e. nitrogen, oxygen, chlorine and bromine which explains why the most electronegative elements are those in the top right hand corner of the Periodic Table.

The actual numbers of the Pauling Electronegativity Scale depend on his calculations. All alternative versions end up with slightly different values but the same trend, all identifying the same elements as the most electronegative. To reinforce the concept and to illustrate why electronegativities decrease moving down a Group it is fruitful to visualise the first two members of Group VII, fluorine and chlorine, and then the first two Group VI elements, oxygen and sulfur. In each case the filled energy levels, the core electrons are highlighted in bold in the linear notation and by rectangles in the *'arrows in boxes'* notation.

		1s	2s	2p	3s	3p	4s	Electronegativity
Group VII *(Effective atomic number +7)*								
$_9$F	$1s^2\,2s^2\,2p^5$	⬆⬇	⬆⬇	⬆⬇ ⬆⬇ ⬆	☐	☐ ☐ ☐	☐	4.0
$_{17}$Cl	$1s^2\,2s^2\,2p^6\,3s^2\,3p^5$	⬆⬇	⬆⬇	⬆⬇ ⬆⬇ ⬆⬇	⬆⬇	⬆⬇ ⬆⬇ ⬆	☐	3.0
Group VI *(Effective atomic number +6)*								
$_8$O	$1s^2\,2s^2\,2p^4$	⬆⬇	⬆⬇	⬆⬇ ⬆ ⬆	☐	☐ ☐ ☐	☐	3.5
$_{16}$S	$1s^2\,2s^2\,2p^6\,3s^2\,3p^4$	⬆⬇	⬆⬇	⬆⬇ ⬆⬇ ⬆⬇	⬆⬇	⬆⬇ ⬆ ⬆	☐	2.5

If we compare:-

- **Fluorine** and **oxygen**: the shielding by the core electrons is the same, as both atoms have a filled first energy level only (two electrons only) but fluorine has the effective atomic number of +7 whilst oxygen has a value of +6. This explains why fluorine is more electronegative than oxygen since the shielding is the same in both cases.

  It also explains why fluorine forms a 1-, *fluoride*, ion and oxygen forms a 2-, *oxide*, ion: both ions acquire the electronic configuration and stability of neon, the next Noble gas.

- **Chlorine** and **sulfur**: again the shielding for both atoms is the same (energy levels 1 and 2 and a total of ten electrons) so can be disregarded as they are both the same. The effective atomic number of chlorine is, however, +7 whilst that of sulfur is +6 hence the electronegativity of chlorine is greater than that of sulfur.

  On reaction, both elements acquire sufficient electrons, one for chlorine and two for sulfur, to acquire the stability of the next Nobel Gas, argon.

The remaining two conclusions now become apparent:
- Fluorine is more electronegative than chlorine since, despite both atoms having the same effective atomic number, fluorine is smaller (having only one set of core, inner, electrons) than chlorine so the nucleus has more pulling power.
- The same principle applies to oxygen and sulfur.

We can also explain why oxygen, in the oxide form, forms more stable ionic compounds than does sulfur in its sulfide form.

Oxygen has a higher *electronegativity* than sulfur since, although both have the same *effective atomic number*, sulfur has an extra, inner core, shell of electrons which shields the outer electrons from the nucleus. This means that the outer electrons of sulfide, S^{2-}, ions are further from the nucleus than are the outer electrons of oxide, O^{2-}, ions and so are less tightly held.

Many textbooks state that there are only four electronegative elements of importance and exclude bromine. This appears contradictory because the electronegativity of bromine is used to explain the hydrobromination of alkenes such as the reaction of ethene with hydrogen bromide:

This can only be explained by the electronegativity of the bromine atom since the above reaction occurs at room temperature whereas the hydrogenation of ethene i.e. reaction of ethene with hydrogen (H_2) requires catalysis and a high temperature to occur.

The concepts of *effective atomic number* and of *electronegativity* go a long way to explaining ionic bonding and the reactivity of many small organic molecules.

We can also explain the reactivity of elements in a particular Group.

Reactivity in groups I and II increases moving down the group. This is because each successive element has one more inner shell of electrons than the one above it. The electrons repel each other and shield outer shells from the nucleus as the electrons also cancel out some of the nuclear charged and this means that the larger the element i.e. the greater the number of inner, core, shells and electrons the less tightly the outer electron is held and the meore easily it is removed by an element such as a halogen.

On the other hand, moving down group VII, reactivity decreases also due to the increasing numebr of inner electrons. To some extent, these repel the incoming electron and makes it harder for the nucleus to grab hold of the electron which, however, it clearly does and so reactivity is a balance between effective nuclear charge, electronegativity and the shielding of the nucleus by the shells of inner electrons

There is one more consequence of these concepts and that is *bond polarity* which is discussed next.

Bond Polarity

When two identical atoms bond as in H_2, Cl_2 etc; neither *effective atomic number* nor *electronegativity* are important since the bonding pair (remember that electronegativity is only concerned with bonding electrons) is shared equally between the atoms because there is no reason for one atom to demand more than its fair share. We can represent this with **x** presenting an electron as follows for the hydrogen and chlorine molecules:

$$H \overset{\text{x}}{\underset{\text{x}}{-\!\!\!-\!\!\!-}} H \qquad Cl \overset{\text{x}}{\underset{\text{x}}{-\!\!\!-\!\!\!-}} Cl$$

Together however the, arithmetic, *effective atomic number* and *electronegativity* can be used to explain, as described previously above:-

1. The existence of ionic bonding due to some atoms stealing electrons and other atoms just letting them go;
2. Why some elements form covalently bonded compounds where the electrons are shared because neither has sufficient power to steal electrons and so simply share them.

By studying numerous compounds it was found that small organic molecules, with low melting and boiling points, contained atoms of similar electronegativity and by comparison, it was determined that any two bonded atoms with a difference of electronegativity of approximately 0.5 or less were in that category. This is just a rule of thumbe but does appear to work. So, for example, all C – H bonds are covalent with the atoms essentially sharing the pair of bonding electrons since the difference between the electronegativities of carbon (2.5) and hydrogen (2.1) is 0.4. Again this was determined by comparisons and not by any sophisticated calculations and these two concepts can also be used to explain why water dissolves ionically – bonded compounds which is discussed now.

Polarity of water

If we compare the electronegativities of oxygen (3.5) and hydrogen (2.1) we find that there is a clear difference of 1.4 which is far greater than the 0.5 difference.

Although the oxygen and the hydrogen atoms in the water, H_2O, molecule are covalently bonded i.e. the oxygen and hydrogen atoms share a bonding pair of electrons they are *not* shared equally. Oxygen, with six times the effective atomic number of hydrogen and hence much more power over the electron, has much more than its fair share of the bonding pair. This means that the oxygen atom has a slight surplus of negative charge but not a full negative charge and we term the molecule as *polar*, with the oxygen being *partially negatively charged*, which is denoted by δ-, whilst the hydrogen atoms are *partially positively charged* and are denoted by δ+. Since these are not full charges, they cannot be quantified and so there is no requirement for the numbers of δ- and δ- symbols to be equal and, hence, we draw the water molecule as shown below:

$$\delta^+ \; H \overset{\delta^-}{\underset{\diagdown \; H \; \delta^+}{\diagup \; O}}$$

The polarity of the bonds, and hence the polarity of the molecule explains the formation of aqueous solutions of ionic compounds due to their solubility and aqueous acids which are discussed next.

Understanding the water solubility of ionic compounds

When dissolving an *ionic* compound the water molecules surround the ions: the cations align with the partially negative oxygen oxygen of the water molecule and the anions are attracted to the partially positive hydrogen atoms are surrounded by the partially positive hydrogen atoms.

The dashed line - - - - indicates a partial bond between the solute ions and the solvent molecules as shown below:

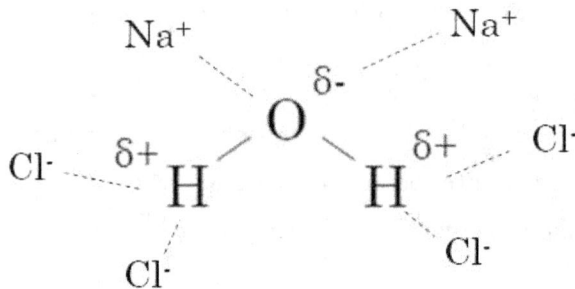

The interaction of the ions with water are not fixed and since ions constantly travel around and through the water and, indeed, the water molecules travel through the liquid itself the partial bonds are formed and broken constantly.

This can be generalised to to explain the solubility of, positively charged, cations and, negatively charged, anions.

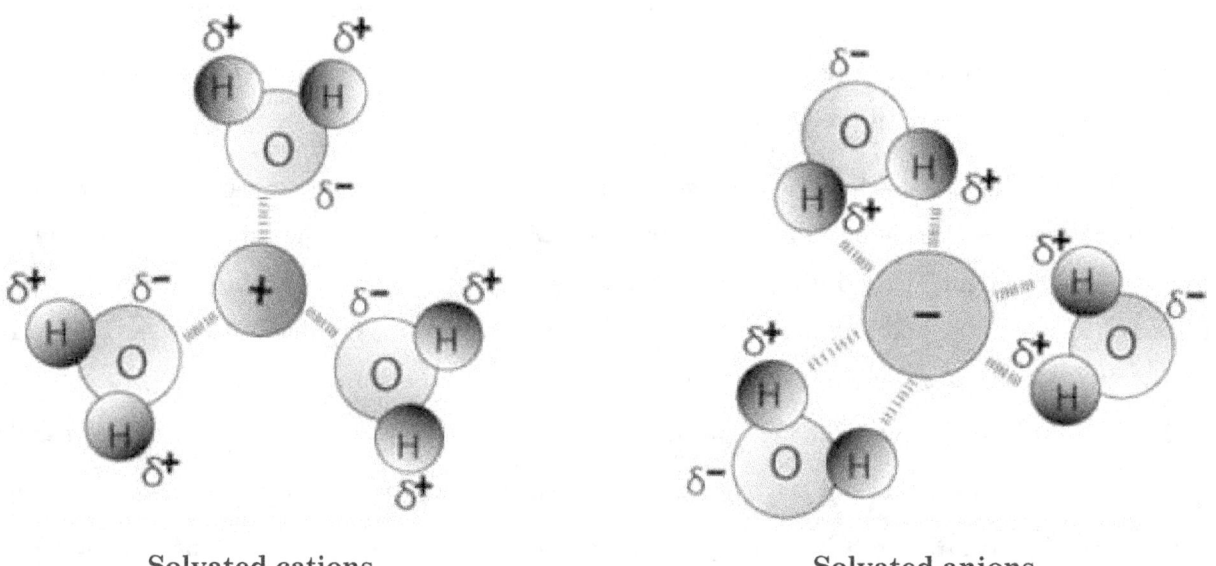

Solvated cations   Solvated anions

These concepts are important in understanding the formation of aqueous acidic solutions and why they only form in water.

Understanding the formation of aqueous mineral acids

There are a number of definitions of an acid but the only one appropriate here is that of an aqueous acid.

All schoolchildren learn about the pH scale and learn that acids have a pH below 7, pure water has a pH of 7 and alkalis have a value above 7 and up to 14. The definition of pH is the minus logarithm (to the base 10) of the concentration of hydrogen ions:

$$pH = - \log_{10} [H^+]$$

where the logarithm, the log, is simply the power to which the base number, in this case, 10, must be raised to calculate the number so, for example, since $100 = 10^2$ then $\log_{10} (1000) = 2$ and the $\log_{10}$ of 1000 is 3 since $1000 = 10^3$. The pH scale is extremely useful since the concentrations of the hydrogen ion are low and the numbers can be difficult to compare. In contrast, the definition of pH gives us a simple scale of 0 – 14 and the minus sign is simply added to keep all values positive.

The scale was devised, in 1909, by Søren Peder Lauritz Sørensen. Sørensen was head of the research laboratory of the Copenhagen Carlsberg factory and created it whilst researching how to ferment lager. This laboratory was also where the, previously mentioned, Kjeldahl technique for determining the nitrogen composition of substances was created. The concept of the hydrogen ion was completely accepted and chemists still use the term despite the fact that, as we now know, the hydrogen ion is simply a proton. The polarity of the water molecule explains the formation of aqueous mineral acids.

Taking the formation of hydrochloric acid as one simple example, there are two steps:
1. The, gaseous, hydrogen chloride is bubbled through water and dissolves.
2. The, polar, hydrogen chloride molecule breaks apart, known as *dissociation*, into separate hydrogen and chloride ions. This confers the stability of argon's electronic configuration on the chlorine atom and now ion, whilst the hydrogen ion is simply a proton.

We can represent the process as shown below:

$$H \xrightarrow{\quad X \quad}_{X} Cl \xrightarrow{\quad H_2O \quad} H^+_{(aq)} + Cl^-_{(aq)}$$

which demonstrates that, after the polar HCl molecule dissolves in water, it breaks apart forming hydrogen (H^+) and chloride (Cl^-) ions both of which are surrounded by water molecules.

The question then arises why some acids are stronger than others i.e. aqueous solutions contain greater concentrations of hydrogen ions. There must be a reason for this otherwise all acids would have the same pH value and the reason lies in the stability of the negative ion, the base. The more stable the negative ion, the more that can exist and hence the greater the concentration of hydrogen ions produced. Many of these stable negative, counter ions, contain electronegative elements as demonstrated by HCl(aq) and which will be demonstrated when we consider the formation of organic acids.

Firstly, however, we need to consider how the hydrogen, H^+, exists in water since as a proton it cannot exist alone and exists as the *hydronium ion* which is considered next.

The hydronium ion

The hydrogen (H^+) ion is a more interesting case since H^+ is simply a proton. In chemical terms the use of the term hydrogen ion remains extremely commonly used since it was termed long before the existence of the proton was discovered. A tiny particle with a whole positive charge, a proton cannot exist on its own since its charge/volume ratio is too great for it to remain isolated. In reality, the hydrogen ion jumps around from water molecule to water molecule and is better written as a hydrogen ion and a water molecule together [H_3O^+] where - - - - - represents a temporary attraction between the hydrogen ion and the water molecule.

$$H_2O \text{ ---------- } H^+ = H_3O^+$$

Since the hydrogen ion, i.e. a proton, has no electrons at all, sharing one electron will not add much to its stability whereas sharing a pair, also referred to as donating a pair of electrons and which can be represented as

$$H_2O: \longrightarrow H^+$$

demonstrates that oxygen confers the stability of the helium electronic configuration (one filled shell) on the hydrogen ion. The pair of electrons is represented by a colon **:** and is referred to as a *lone pair of electrons* or simply as a *lone pair*. The arrow describes the donation of the lone pair to to the hydrogen ion.

This is analogous to the dissolution of any positive ions as discussed earlier when considering sodium chloride and can be visualised as shown below:

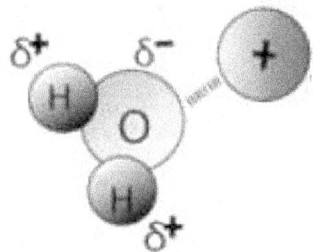

We can understand this by considering the electronic configuration of oxygen:

Two of the 2p orbitals contain one electron each and they are the ones which become filled when bonding with, for example, hydrogen who donates its electron to the orbitals explaining, as discussed before, why water has the formula H_2O. This leaves one pair of 2p electrons not used in covalently bonding to other atoms. This pair is highlighted

and this is the *lone pair* which stabilises the hydrogen ion (H^+).

Understanding the formation of aqueous organic acids

These concepts explain the acidity of organic acids which are usually weak, typically with a pH value between 3 and 5. The polarity of the water molecule also explains water's ability to dissolve materials such as ethanoic acid (CH_3COOH) which is itself polar due to the presence of the oxygen atoms and the existence of the $O - H$ bond.

$$CH_3CO_2H \xrightarrow{\quad H_2O \quad} H^+_{(aq)} + CH_3CO_2{^-}_{(aq)}$$

and can also be written as:

$$CH_3CO_2H(s) + H_2O \longrightarrow H_3O^+(aq) + CH_3CO_2{^-}(aq)$$

The negatively charged, counter ion, $CH_3CO_2{^-}$, is stabilised due to the existence of two electronegative oxygen atoms and the negative charge can be spread between them and can be written as two, so called, resonant structures as shown below, where $\longleftrightarrow$ indicates equivalent structures.

Understanding the formation of aqueous alkali solutions

Ammonia, NH_3, is a gas which when bubbled through water, dissolves and breaks the water molecule apart, forming ammonium, NH_4^+, and hydroxide (OH^-) ions. The presence of the aqueous hydroxide ions explains why the solution is a strong alkali with a pH between 11 and 12.

$$NH_3\ (g) + H_2O(l) \longrightarrow NH_4^+(aq) + OH^-(aq)$$

This also explains why acids can be titrated with aqueous ammonia since the acid supplies the hydrogen [H+] ions and ammonia through the above equation supplies the hydroxide (OH^-). Writing the aqueous ammonia formula as NH_4OH, we have:-

$$HCl_{(aq)} + NH_4OH_{(aq)} \longrightarrow NH_4Cl(aq) + H_2O(l)$$

On evaporating the water, the product, ammonium chloride (NH_4Cl) is produced as a white, crystalline solid resembling sodium chloride. This also demonstrates that the overall neutralisation process is:-

$$H^+_{(aq)} + OH^-{_{(aq)}} \longrightarrow H_2O_{(l)}$$

We can now understand the solubility of small chain alcohols and the acidity of compounds which all contain the – COOH functional group. No alcohols, ketone or aldehydes create acidic solutions but those containing the –COOH group do and this is easily explained.

We will consider alcohols first.

Understanding alcohols

Alcohols contain a C – O – H group which is polar. This means that the partially charged oxygen (δ-) and hydrogen (δ+) atoms interact with the polar water molecule as discussed earlier. The solubility of alcohols decreases with increasing carbon chain length since the long, non – polar, carbon chains get in the way and reduce these interactions.

It turns out that the ideal combination of polarity and chain length is that of ethanol which can mix with water in all concentrations to form a homogenous solution.

$$
\begin{array}{ccc}
& H & H \\
& | & | \\
H- & C- & C-O-H \\
& | & | \\
& H & H
\end{array}
$$

Ethanol does not form an acidic solution however. In other words, the hydrogen atom does not break off to form a H⁺ ion and this is simply because the remaining negative charge would have to exist on the residual, ionic, fragment, effectively on the single oxygen atom since the carbon atoms have no spare orbitals to hold the electron:

$$
\begin{array}{ccc}
& H & H \\
& | & | \\
H- & C- & C-O^{-} \\
& | & | \\
& H & H
\end{array}
$$

Neither carbon nor hydrogen are sufficiently electronegative to accept this charge and it would have to reside on the oxygen atom which would be tempted to recombine with a hydrogen ion. That these does, indeed, occur is deomnstrated by the fact that all aqueous solutions of those alcohols which are water soluble are neutral i.e. they have a pH of 7 and do not undergo neutralisation reactions with either acis or alkalis.

The solubility of alcohols varies depending on the number of carbon atoms in the carbon chain.

Methanol, CH_3OH, ethanol, CH_3CH_2OH and 1-propanol, $CH_3CH_2CH_2OH$ are all infinitely soluble in water i.e. any quantity will dissolve. The solubility of the longer chain alcohols decreases rapidly with increasing chain length. For example, the solubility of 1-butanol, $CH_3CH_2CH_2CH_2OH$ is 9g / 100cm³ of water, that of pentanol, $CH_3CH_2CH_2CH_2CH_2OH$ is 2.7g / 100cm³ of water.

Longer chain alcohols are effectively insoluble. Since the C – O – H bond is the same in each case the changes in solubility must be due to the increasing chain length which is non-polar and cannot interact with the polar water molecules.

This is demonstrated by the differences in the solubility of the four isomers of butanol. Three have relatively low solubilities in water but one, that with the shortest carbon chain, shown right, has infinite solubility.

$$
\begin{array}{c}
CH_3 \\
| \\
H_3C-\!\!-\!\!-OH \\
| \\
CH_3
\end{array}
$$

We can now consider the solubility of carboxylic acids but to do that we need to compare all carbonyl - containing compounds.

Understanding Carbonyl compounds

All carbonyl compounds contain a C = O bond and there are five classes that are pertinent to this volume, aldehydes, ketones, carboxylic acids & esters and acid chlorides. Esters also contain a single C – O bond and the classes are discussed below.

Pertinent to all classes are the electronegativities of the elements present. The electronegativity of carbon is 2.5 whilst that of oxygen is 3.5. Even though the bonding is covalent i.e. the electron pair is shared by the carbon and oxygen atoms they are **not** shared equally. Oxygen has more of the shared pair than carbon and so the oxygen is said to be *partially negatively charged* and the carbon atom, which is now said to be *electron – deficient*, is *partially positively charged* and the bonds are described as *polar* and can be written as:-

$$\overset{\delta+}{C} - \overset{\delta-}{O} \qquad\qquad \overset{\delta+}{C} = \overset{\delta-}{O}$$

The δ symbol in the notation signifies that it represents an unquantified amount and so this means that, unlike in ionic compound where the number of positive and negative charges must be equal, this does not have to be the case here. They are described as *polar bonds* to distinguish them from ionic bonds.

In the following examples, functional groups in the generic formulas are represented by R, R^1 etc; and can be any carbon containing groups.

Aldehydes contain a hydrogen atom attached to this carbon atom and can be written as

$$\begin{array}{c} H \\ | \\ R - C = O \end{array}$$

Ketones do not have a hydrogen attached to that carbon atom which is bonded to two other carbon atoms

$$\begin{array}{c} R^1 \\ | \\ R - C = O \end{array}$$

However large the molecule, these functional groups always exist on a terminal i.e. end carbon atom. The small aldehydes nor ketones are freely soluble in water but their solubility decreases with increasing carbon chain length for the reasons given above for alcohols. In addition, since they do not have an O – H group, they cannot form acidic solutions.

Carboxylic acids are a different matter though since they do possess a hydroxyl (–OH) group as demonstrated by the following generic formula:-

$$\begin{array}{c} OH \\ | \\ R - C = O \end{array}$$

As explained earlier, when they dissolve in water, the hydrogen can break off from the hydroxyl group forming a hydronium ion and a stable, negatively charged, counter ion which can be written as

$$\begin{array}{ccc}
O^- & & O \\
| & \longleftrightarrow & \| \\
R - C = O & & R - C - O^-
\end{array}$$

The compounds form stable acidic solutions since the remaining, negatively counter ion, is stabilised by the presence of the two electronegative oxygen atoms.

All carboxylic acids with one to four carbon atoms are infinitely soluble in water but the solubility then decreases rapidly with increasing carbon chain length for the reasons we have discussed ealier.

Esters, which are formed through the reaction of a carboxylic acid and an alcohol, have the generic formula:

$$\begin{array}{c}
OR^1 \\
| \\
R - C = O
\end{array}$$

where, again, R and R^1 represent any grouping containing a carbon atom bonded to the carbonyl group.

Small chain esters dissolve in water and react very slowly with the solvent. If acid or base catalysed the ester can be hydrolysed i.e. reacts with water to form the carboxylic acid.

We now understand why small chain carboxylic acids dissolve in water and why they form acidic solutions and also understand why small alcohols, aldehydes and ketones dissolve in water but only form neutral solutions whilst the solubility decrease with relative molecular mass i.e. increasing chain length.

Acid (acyl) **chlorides** have the generic formula RCOCl but the only important one is ethanoyl chloride.

$$\begin{array}{c}
Cl \\
| \\
H_3C - C = O
\end{array}$$

They do not dissolve in water instead reacting extremely violently with water producing the corresponding carboxylic acid and hydrogen chloride gas.

We now know enough to explain, understand and apply the principles of infra red and mass spectrometric analytical techniques and then multi-nuclear magnetic resonance.

These will discussed in order.

Chapter VI

Infrared spectroscopy and mass spectrometry

This chapter considers the:-

- Theory behind infrared spectroscopy;
- Application of infrared spectroscopy in more detail with three examples
- Mass Spectrometry of organic compounds

Infrared spectroscopy reveals the presence or absence of specific bonds such as the $C = C$, $C = O$ and $O - H$ bonds and also reveals the presence or absence of specific functional groups such as the carboxylic acid functional group which contains both a $C = O$ and a $C - O - H$ bond.

Mass spectrometry also reveals the presence of specific fragments and can reveal much useful information about the structure of a molecule.

Both techniques are discussed now.

Principles of Infrared Spectroscopy

Infrared spectroscopy, which was first invented in the 1940s, is an essential tool in the determination of the presence, and equally importantly the absence, of functional groups.

Whilst microwaves can make molecules rotate and of are great use in air pollution analysis, infrared spectroscopy involves the impact of infrared radiation on the stretching and twisting etc; of molecular bonds. The radiation increases the length of the bond without actually breaking it and is part of the electromagnetic spectrum known as non – ionising radiation (all regions with less energy i.e. lower frequency than x – rays).

An effective way to understand how infrared radiation impacts on a molecule is to stand up, legs apart and arms pointing above, and at 45°, to the shoulders, as if performing a star jump. The waist represents the central atom, arms and legs depict covalent bonds and the hands and feet represent the other bonding atoms.

- Twist from the waist so that the arms and hands are now at 90° to the body and this is, unsurprisingly, described as a *twist*.
- Standing straight again with hands close to the shoulders: Extend both arms, stretching as much as possible. This is a *symmetrical stretch*.
- Standing straight again with hands close to the shoulders, point with one hand as if landing a punch. Since one hand is as far from the body as possible whilst the other is close to the body this is an *asymmetrical stretch*.
- Arms in the air again pointing as far apart as possible but in line with the body and clap your hands together above your head. This motion is *scissoring*.
- Again standing straight put both arms in the air again, like a star jump move. Move, simultaneously, both arms from left to right or vice versa. This is *rocking*.

In molecular terms, some of these movements are illustrated below.

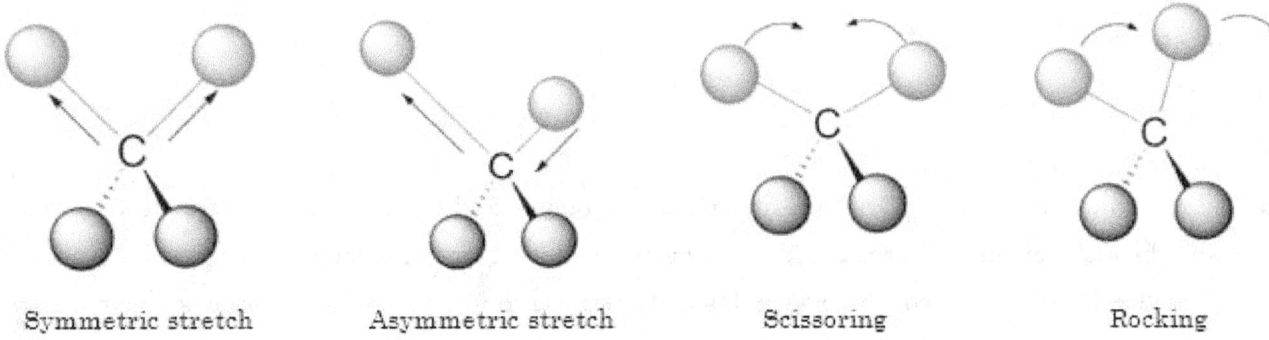

Symmetric stretch   Asymmetric stretch   Scissoring   Rocking

In these pictures a plain straight line links atoms in the plane of the paper, the hatched line indicates atoms behind the paper and the dark arrow indicates those in front of the pagae and are the standard way of displaying a three dimensional molecule in two dimensons.

When molecular atoms perform these actions many of the absorptions appear in the region below 1500 cm^{-1}, and can overlap making that part of the spectrum extremely difficult to interpret. The combination of peaks is important and unique to the specific compound and the specific wavenumbers are affected by the presence of isotopes since the differing masses affect the necessary wavenumbers. Since, however, they are unique to the compound they are immensely useful for comparison purposes and hence it is known as the *fingerprint region*.

The Effects of Bond Strength and Electronegativity

The two principle reasons for the specific frequencies at which stretches occur within the infrared region are the masses of the two atoms and the bond strength. The *bond strength* depends on the atoms present and can be defined as the energy required to break the bond. It is also known as the *bond energy* and the *bond enthalpy* and is measured in kJ mol^{-1}. The term *bond strength* is used throughout this text as it is the most easily understood.

It is also possible to compare bond strengths by comparing bond lengths. The bond length is the distance between the nuclei of the two atoms and is measured in picometres (pm) which is equivalent to 10^{-12} m. The bond length is longer with a weaker bond, since the bonding electrons are less able to hold the atoms together and shorter for atoms held more tightly together. The other hugely important factor is the *electronegativity* of the bonding atoms. Bond strengths and electronegativity are discussed below.

Bond strength

Bond strengths, i.e. the energy required to break bonds can be compared by measuring the distance between the two atoms involved in the bond. For our purposes, the most instructive comparison is that of the bond lengths (the distances between the carbon atoms in the so called carbon – carbon single, double and triple bonds as shown below and becomes informative when we recall that the stronger the bond the shorter the bond length.

Molecule	ethane	ethene	ethyne
Bond length (pm)	154	133	120
Wavenumber (cm^{-1}) typical	1200	1650	2200

As discussed earlier, the concept of carbon – carbon single, double and triple bonds arose from the concept of valency and the requirement for carbon to form four bonds and hydrogen to form just one. It is clear from the bond lengths that a carbon = carbon double bond is **not** twice as strong as a carbon – carbon single bond since if so it would be half the length of a C – C bond and it is not. The same observation applies to the triple carbon – carbon bond. It is also noticeable that all the C – H bonds are of the same length and hence of the same strength. This is significant for clear chemical purposes but it also indicates that the nature of one non-polar bond has little if any effect on other bonds formed by the same atom. It is different for polar molecules which are discussed below when we also consider the relevant parts of the concept of electronegativity on spectroscopy. Finally, with reference to alkanes, alkenes and alkynes the carbon – carbon single, double and triple bonds the bond strength differences means that the energy required to stretch, twist, rock etc; the atoms explain why there are different absorbances for all three bonds and also explains why all C – H bonds appear in the same region.

Electronegativity

This concept has been discussed before but some concepts are worth reconsidering since they have different consequences and some are too difficult to only consider once and understanding increases with reiteration. It is also worth mentioning here that the fact that many atomic concepts are difficult to understand can be daunting to one new to the subject. There is nothing wrong with that since all of us have experienced the same difficulties both as pupils and the scientists who struggled to understand the meaning of their results. The only people who find these concepts trivial on introduction are those who not understand them at all.

Electronegativity is defined as *the ability of an atom to attract both of the bonding pair of electrons in a covalent bond towards itself*. The most electronegative atoms are nitrogen, oxygen, fluorine, chlorine and bromine and this is explained by calculation of the *effective atomic number* of the atoms.

The halogens all have an effective atomic number of seven whilst that of oxygen is six. The relative atomic number of hydrogen is simply one which means that oxygen has six times more power than hydrogen to attract the bonding pair whilst the halogens are even stronger and the bonds are even more distorted in terms of electron sharing.

The presence of electronegative elements causes a distortion in the distribution of the bonding electrons in a covalent bond as demonstrated by comparing the bonding distribution in the hydrogen molecule, the hydrogen chloride molecule and the water molecule.

The *hydrogen* molecule, H_2, contains two identical atoms obviously of the same electronegativity so the bonding pair can be visualised as in the middle of the bond where **x** represents one electron.

$$H \underset{\text{x}}{\overset{\text{x}}{\rule{3cm}{0.4pt}}} H$$

Chlorine is the third most electronegative element so so pulls the bonding pair of electrons towards itself as shown below in the *hydrogen chloride* molecule:

$$H \underset{\text{x}}{\overset{\text{x}}{\rule{3cm}{0.4pt}}} Cl$$

Since the hydrogen atom is deshielded and possesses less of the bonding pair than the chlorine, it is described as *electron deficient* and is represented by δ+ whilst the chlorine atom, with more than its fair share of the bonding pair and so *electron rich* is represented by δ- so we can now draw the, *polar*, HCl molecule as:

$$H \overset{\delta+}{\rule{2cm}{0.4pt}} Cl^{\delta-}$$

In these commonly used notations the term δ+ and δ– are not quantified, numerical values and simply signifies that the atom is partially charged.

The effect of electronegativity is clearly demonstrated by comparing the absorbances of the C – H and C – Cl which typically occur at ~2950 cm^{-1} and ~750 cm-1.

Infrared Spectroscopy in more detail

Infrared radiation causes the bond in a molecule to stretch (symmetrically and asymmetrically), rotate, rock or twist.

No molecule is ever at rest since at even the lowest temperatures it is still rotating but when electromagnetic radiation of slightly higher energy, in this case ir wavelengths, is absorbed, the molecule absorbs a portion of it, making use of it to stretch, twist, rock, scissor etc;

The radiation is then re-emitted when the bond releases the energy and the atoms return to their usual place. This is called *relaxation*. Whilst the ir radiation comes from one specific direction the emission of energy on relaxation occurs in all directions and effectively none of the absorbed and then emitted energy reaches the detector. This means that the molecule absorbs radiation and that this absorbed radiation does not reach the detector. In the actual experiment, if the compound is dissolved in a solvent, the ir beam is split through one reference container, containing the solvent only, and an identical container of the dissolved sample material. The difference in the received light represents the absorption by the sample.

The schematic arrangement of an infrared experiment is here.

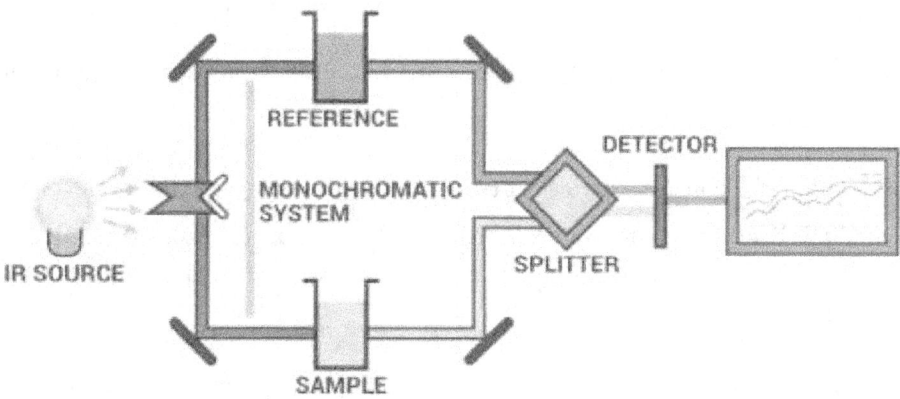

Initially, an infrared spectrum can be daunting and appear incomprehensible however when the regions are broken down analysis can become much simpler than at first appears.

As explained above bonds can stretch, symmetrically and asymmetrically, and the bonded atoms can rock, scissor, twist etc; Each of these stretches requires a specific frequency of radiation. It is possible to calculate the frequency at which each of these will occur but that is unnecessary for structural analysis and is rarely performed by chemists.

Instead we look for the presence of specific functional groups whose presence can be confirmed by absorptions in specific regions and their absence can be completely ruled out by the absence of such peaks. This is made simpler by the fact that, irrespective of the structure of the molecule, particular stretches *always* occur in specific regions of the spectrum and can be identified. For example, a carboxylic acid will contain a – COOH group and will exhibit C = O, C – O and O – H peaks.

For our purposes the groups to look for are:

* O – H * N – H * S – H * C = O * C = C * C – O

The first issue to be addressed is that of the unit, the *wavenumber*. The wavelengths or frequencies at which molecules absorb ir radiation can be very large or very small numbers respectively. This makes it difficult to compare them and so instead we use the concept of the wavenumber which has the unit cm^{-1}. The wavenumber is simply the number of waves present per centimetre and as such is a frequency. The frequency range now comprises numbers ranging between 4000 and ~500 cm^{-1}, which are much more comprehensible. The reason why spectra stop at 500 cm^{-1} is for practical reasons only. The solid samples are dispersed with a paraffin wax known as nujol and contained between discs made of sodium chloride whilst solutions are contained within containers with sodium chloride or, more usually, potassium bromide windows. These materials are transparent to ir radiation between 4000 and 500 cm^{-1} but absorb all of it below 500 cm^{-1}.

There are specific regions to examine which may be used to identify the *presence* or the definite *absence* of particular functional groups. The most important are identified in this table and includes the $C \equiv N$ (nitrile) bond which has not been discussed to date.

Group	Wavenumber range (cm^{-1})	Group	Wavenumber range (cm^{-1})
O – H	~ 3500	C = O	~ 1750
N – H	~ 3500	C = C	~ 1650
S – H	~ 3500	C – O	~ 1450
C – H	3000 – 2900	C ≡ N	~ 2250

To summarise, the first three bonds occur at wavenumbers centred on 3500 cm-1, carbonyl (C=O) peaks appear at around 1750cm^{-1}, C=C double bonds absorb at ~1650cm^{-1} and C-O stretches appear at around 1450 cm^{-1} so for structural determination purposes we only need to look at regions centred on 3500, 1750, 1650 and 1450 cm^{-1}. Since all organic compounds contain C – H bonds the region to the right of 3000 cm^{-1} is useless for structural purposes since those peaks will always be there.

The matter is simplified even more if there are for example, oxygen atoms present but no sulfur or nitrogen atoms and it is made even easier using a correlation chart. The specific regions assignable to particular groups did not arise from complicated calculations; rather they were determined by the measuring the spectra of numerous examples of the same classes of compounds and assembling a correlation chart as shown above.

There are two other characteristics of an ir spectrum:

- Most of the stretches appears below 1500 cm^{-1} and cannot be assigned specifically but are a combination of peaks unique to that compound and, if a pure sample is available, the identity of an sample can be assessed by comparison hence this region is known as the *fingerprint region*.
- The peaks can be very broad (O – H), very strong (C = O and C = C) or can be very weak and everything else in between. Peaks are often described as the wavenumber with st, br, w, vw in brackets afterwards representing strong (st), broad (br), w (weak), very weak (vw) etc;

The presence or absence of these bonds, used in the analyses of the compounds in this series are of great significance in the determination of the functional groups that are present or absent and were of enormous importance in chemical advances in the 1950s and 1960s. It is easiest to understand this by actually looking at real spectra. We will look at three: a 3-carbon alcohol (**propanol**), **propanoic acid** and **cyclohexene**.

Propanol

The 3-carbon alcohol, **propanol**, has elemental composition C: 59.92% H: 13.45% O: 26.63% and a relative molecular mass (M_r) of 59.07 g mol^{-1}. The molecular formula is, therefore, C_3H_8O.

Since there is only one oxygen atom present the compound cannot be a carboxylic acid (which requires two oxygen atoms) and other possible classes could be aldehydes, ketones or alcohols. The only three carbon compound aldehyde, propanal, would have the formula CH_3CH_2CHO whilst the only possible three carbon ketone, propanone, will have the formula $CH_3C(O)CH_3$. This only leaves an alcohol and there are two possible structures (isomers) for propanol:

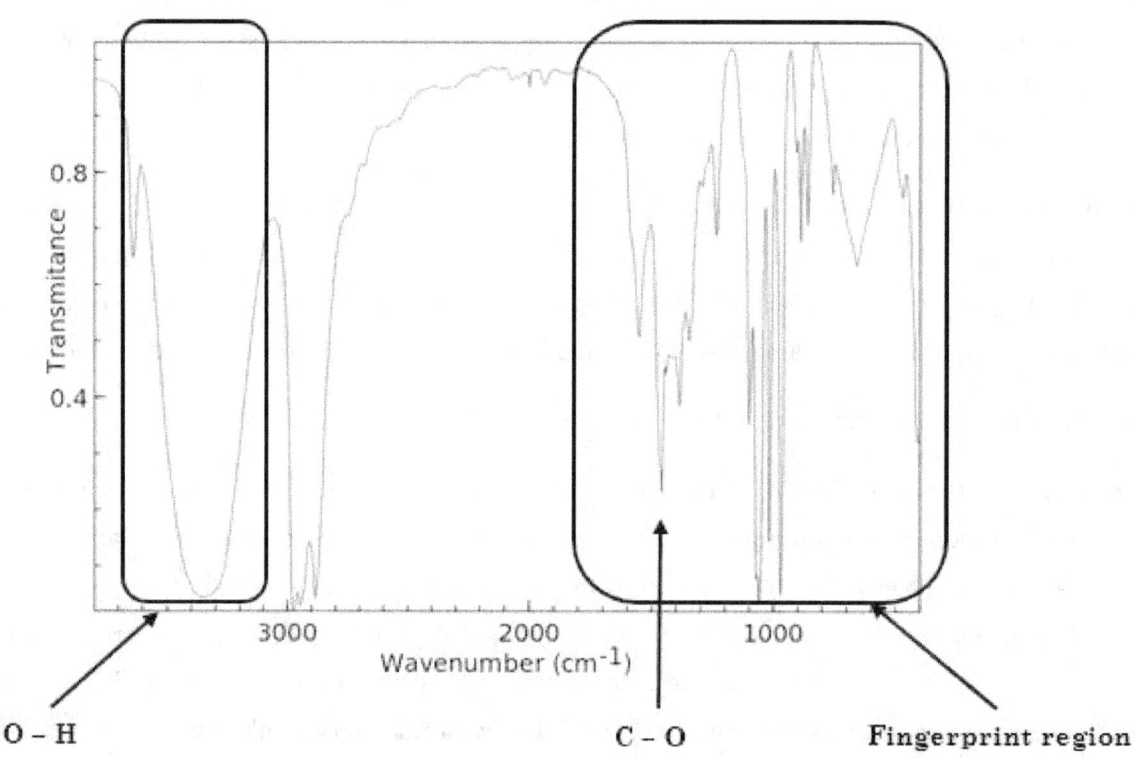

1 – propanol   **2 – propanol**

Remember that there cannot be a third isomer with the –OH group on the left most carbon as that is the same as 1 – propanol. which would be systematically named by the number of the carbon containing the O – H functional group as shown above.

The infrared spectrum shows the presence of C – H, O – H, C – C and C – O bonds but cannot differentiate between the two feasible structures from first principles.

Propanoic acid

Propanoic acid has the elemental composition: C: 48.61% H: 8.18% O: 43.21% and a relative molecular mass of 74.06 g mol^{-1} indicating that the molecular formula is $C_3H_6O_2$.

It is known to be acidic because a dilute aqueous solution has a pH of ~ 3 and it reacts with aqueous alkaline solutions.

Since a carboxylic acid has a –COOH group which from the number of bonds can only exist on the terminal carbon, the only feasible structure is

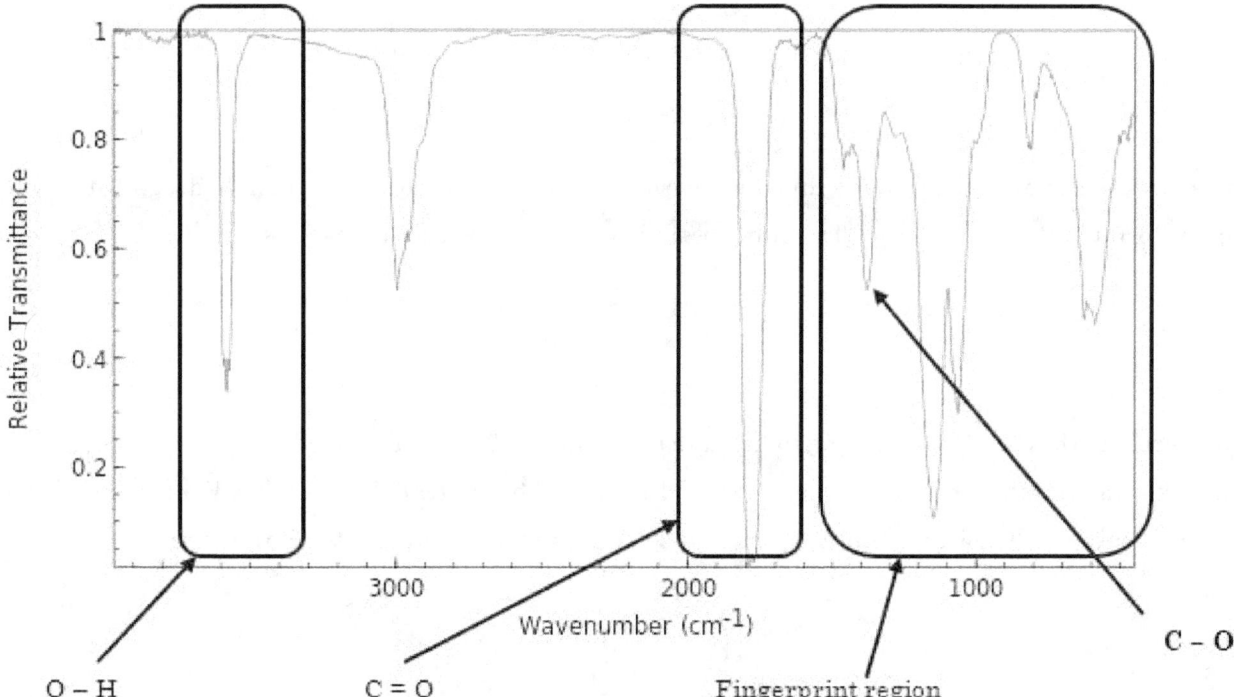

The infrared spectrum supports this suggestion since it demonstrates the existence of both an O – H bond and a C = O bond as well as a C – O bond as highlighted.

The C – O peak at ~1450 cm^{-1} would be absent if the oxygen containing compound was a ketone or an aldehyde since neither class contains a single C – O bond.

Cyclohexene

Cyclohexene has the molecular formula C_6H_{10} and a relative molecular mass of 82.10 g mol^{-1}. That it has a $C = C$ bond is indicated by the traditional test, decolourisation of bromine water and this is confirmed by the presence of a strong peak at ~1650cm^{-1} as highlighted in its infra red spectrum:

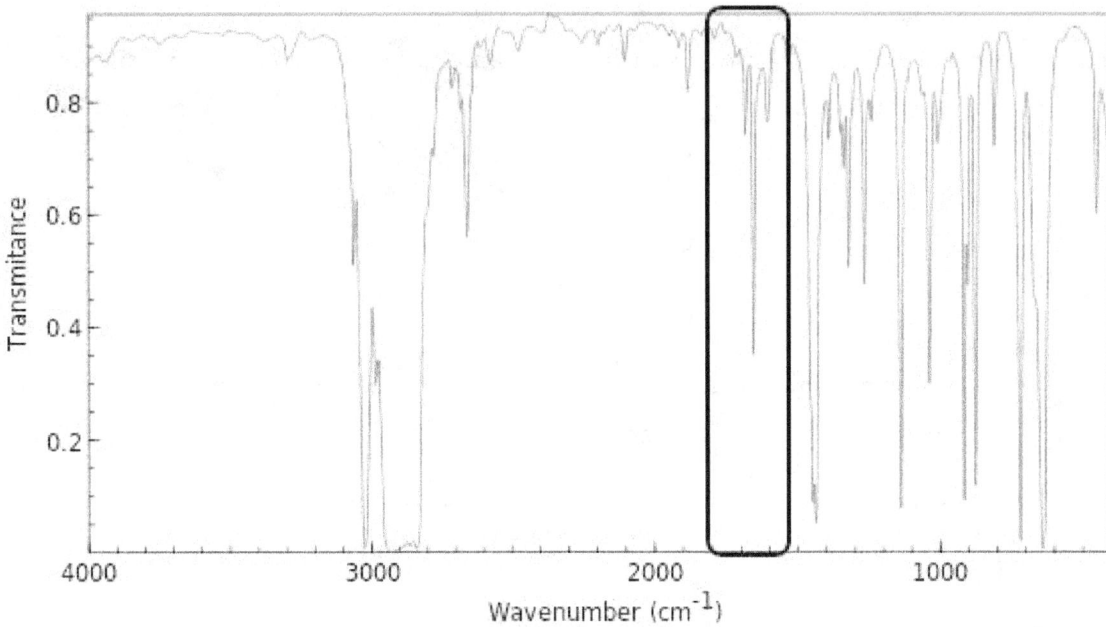

Alkenes can be linear or cyclical but a linear six-carbon alkene would require thirteen hydrogen atoms i.e. the molecular formula would be C_6H_{13}. The only possibility is a cyclic structure as illustrated

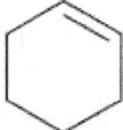

No numbering of the carbons in the ring to indicate the position of the double bond is necessary since wherever we place it, simple rotation of the molecule would lead to the illustrated molecule. For example, the four molecules illustrated below are identical since each can be simply rotated in to all the others.

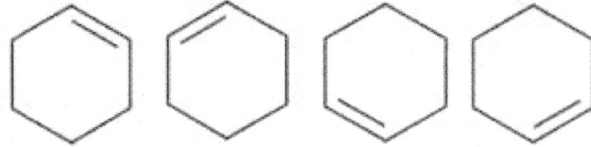

Although infrared spectroscopy is an extremely powerful tool for identifying the presence of functional groups, it has limitations when attempting to distinguish between theoretical isomers.

Another limitation is that although the experiment records the presence of absence of functional groups it does not quantify the number of such bonds. It reveals little information about the carbon – carbon backbone, and nor does it indicate and this nor completely justify the existence of each group in the molecule and this is where mass spectrometry and multinuclear nmr is of use.

Organic Mass Spectrometry

Mass spectrometry has already been discussed in relation to its application and role in the discovery of the isotopes of an atom but it is even more useful in determining the structural composition of organic molecules.

Essentially the mass spectrometric process can be described in a series of steps:

1. **Vaporisation** of the sample which can only be analysed in the gas phase. It is also essential that the entire process is conducted in evacuated apparatus since, otherwise, the charged fragments would be absorbed by the atmosphere present. Vaporisation is essential otherwise the ionisation process (step 2) would merely result in heating the sample and vaporise it anyway.

2. **Ionisation** of the sample.

   This is effectively throwing something at the molecule [M] to knock out an electron and is analogous to throwing a rock at a vase and analysing the broken pieces to see what the vase was made up of − Don't do that!

   The resulting molecular ion [M+] is inherently unstable and, under the exact same conditions, will break apart in exactly the same way every time the experiment is conducted. There are various ways to ionise the molecule. The two most important are:

   1. Throw electrons at it hoping that at least one electron will knock one out of the molecule. This is known as electronic ionisation.

   2. Throw small molecules such as ammonia (NH_3) at the molecule. The ammonia molecules will do to two things: eject an electron and then also bind to the target molecule. This creates an [M+NH_3]+ molecular ion. Naturally the presence of ammonia must be taken into account when studying the fragments.

   Electron ionisation is by far the most commonly used, and useful, technique and is the only one we will consider here. In principle, a stream of electrons impinging on a target molecule could knock out more than one electron thus creating [M^{2+}], [M^{3+}] ions etc; In actuality although such events do occur they do so very rarely and so rarely that they can be ignored.

3. The molecular ion [M+] is then attracted towards the negative plate of an electric field and races towards it, passing through a hole in the plate. This is **acceleration**.

4. It then carries on to pass through a magnetic field. As discussed previously, a moving charge of whatever type will feel the presence of the magnetic field and its path will be altered depending on its mass/charge ratio (m/z). Such a change of path is known as **deflection** and causes spearation of the peaks.

5. The fragments all change their paths very slightly but sufficiently so that they can be **detected** separately. The original measurements were made by using a photographic plate and a series of dots were produced. The lightness or darkness of these spots indicate the relative proportions of each fragments as a total of the whole. Nowadays the detector measures the electrical current generated at each m/z value.

Note: it is important to appreciate that the molecular ion is, by definition, unstable and will produce charged and neutral fragments. The neutral fragments are unaffected by an electric field and play no further part in the experiment.

A schematic diagram of the mass spectrometry process is shown below.

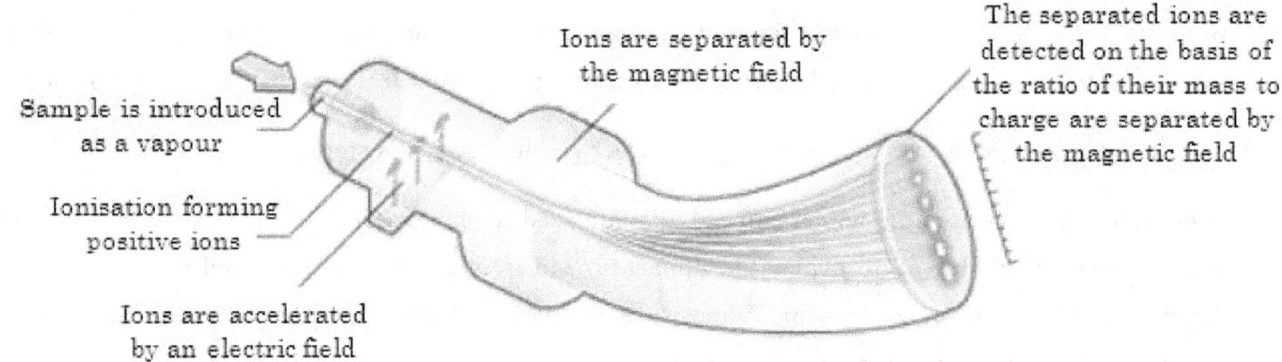

Ions are separated by the magnetic field

The separated ions are detected on the basis of the ratio of their mass to charge are separated by the magnetic field

Sample is introduced as a vapour

Ionisation forming positive ions

Ions are accelerated by an electric field

The mass spectrometry process

The charged fragments are, themselves, also unstable and can fragment as they travel producing further fragments which can produce further fragments. Some can also rearrange themselves into further charged species which can be detected. This can make the final spectrum of very large molecules extremely difficult to interpret but the task is worthwhile since it reveals much about the structure of molecules. All of this occurs within less than one second which is the typical time of flight of the molecular ion and its fragments from ionisation to detection. Such rearrangements are, however, beyond the scope of this introductory text.

If we consider an alkane of molecular formula C_6H_{14}.

The infra red spectrum of the molecule would be of limited value since it would only demonstrate the presence of $C - C$ and $C - H$ bonds together with the absence of a $C = C$ bond. This molecule provides a very clear example of the benefits of the mass spectrometric analysis and the mass spectrum is displayed below.

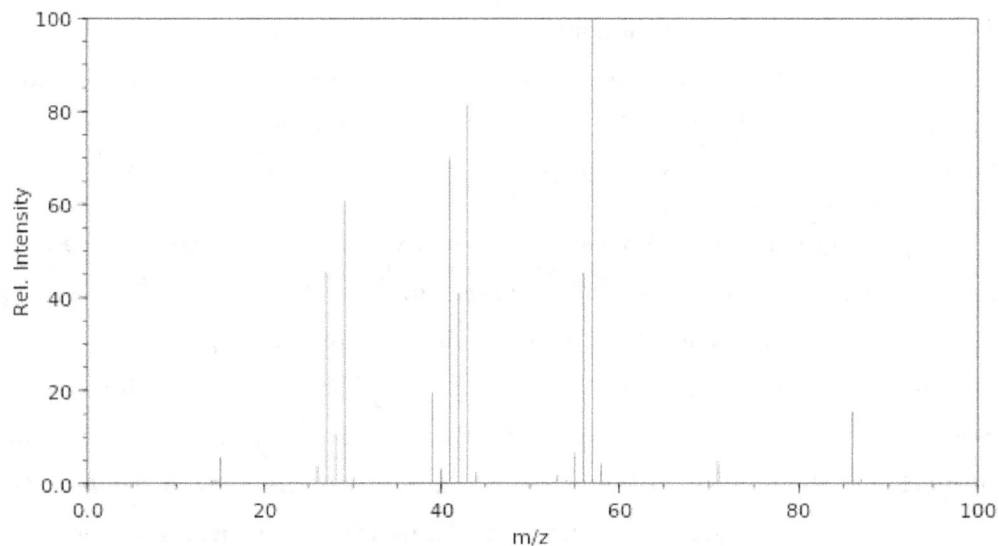

Before we analyse the spectrum in depth there are two points to note.

▪ The analysis of the sample is *not* quantitative by mass and the peaks are scaled such that the most intense is set to be 100% and all other peaks are then of relative height and intensity to that peak. The 100% peak is known as the *base peak*.

- The molecular ion [M]⁺ is *not* the peak with the greatest mass/charge (m/z) ratio. There is always one peak one mass unit greater, the so called [M+1] peak which is always approximately 1% of the height of the molecular ion. This is due to the existence of isotopes of carbon.

The predominant carbon isotope (^{12}C) comprises approximately 99% of all carbon atoms. There are other carbon isotopes, notably ^{13}C which comprises ~1% of all carbon atoms and hence produces a peak one m/z unit higher than the molecular ion but which is ignored in spectrometric analysis. For every molecular ion fragment, however, there is also one peak one unit higher at 1% of the height of the major fragments. They can be ignored but their presence does account for numerous lines in the spectrum which are not determined.

Returning to the above mass spectrum,

Hexane has the molecular formula C_6H_{14} and so has a relative molecular mass (M_r) of 86. The molecular ion [M⁺] has exactly this mass and it is clear that there is a tiny line one unit to the right due to the presence of ^{13}C in 1% of the total number of carbon atoms in the molecules.

There are two straightforward ways to determine the structure of particular fragments:

- The first looks at the lighter fragments and frequently pays dividends as there are few fragments which can form light fragments and there are a number of established rules:
  - A line at m/z = 15 can only be due to a – CH_3 group, that at m/z of 29 can only be due to a – C_2H_5 group, m/z = 43 indicates a – C_3H_7 group etc;
  - A line at m/z = 12 is just a ^{12}C atom whilst a line at m/z = 16 is oxygen and so on.
  - Two peaks at m/z of 35 and 37 in a relative intensity of 3:1 can only be due to the presence of a chlorine atom (caused by the ^{35}Cl and ^{37}Cl atoms which naturally exist in the proportion 3:1) whilst lines of equal height at m/z of 79 and 81 can only due assigned to a bromine atom of which two isotopes exist, ^{79}Br and ^{81}Br and which exist in roughly equal proportions. Naturally, the presence of a halogen atom would already be known from the elemental composition and from the specific tests for the presence of such atoms.
- The second way is a brute force method which is appropriate when a particular structure is suspected is a brute force method of determining the potential fragments. Essentially this involves drawing out the structure of the molecule and drawing lines across which bonds could be broken and then checking if these appear in the mass spectrum.

For example, hexane can be broken, theoretically, as shown below by the dashed lines.

| m/z: | 15 | 71 | | 29 | 57 | | 43 | 43 |

If the molecule has the structure proposed and fragments as suggested then we should expect to see peaks at m/z = 15, 29, 43, 57, 71 as indeed we do confirming the likelihood of this structure. The structure is confirmed later when we investigate ^1H and ^{13}C nuclear magnetic resonance (NMR).

Propan-1-ol or Propan-2-ol?

Returning to the three carbon molecule, **propanol**, which has elemental composition C_3H_8O and a relative molecular mass of 60.08 g mol^{-1}. Propanol has two isomers are possible as illustrated previously but also, for different purposes demonstrated again below.

We know from the ir spectrum that the molecule contains both an $O - H$ and a $C - O$ bond but no $C = O$ bond.

The infra red spectrum could not, however, determine if the $O - H$ functional group exists on the end or middle carbon atom. Unfortunately, mass spectrometry does not really help either as demonstrated below.

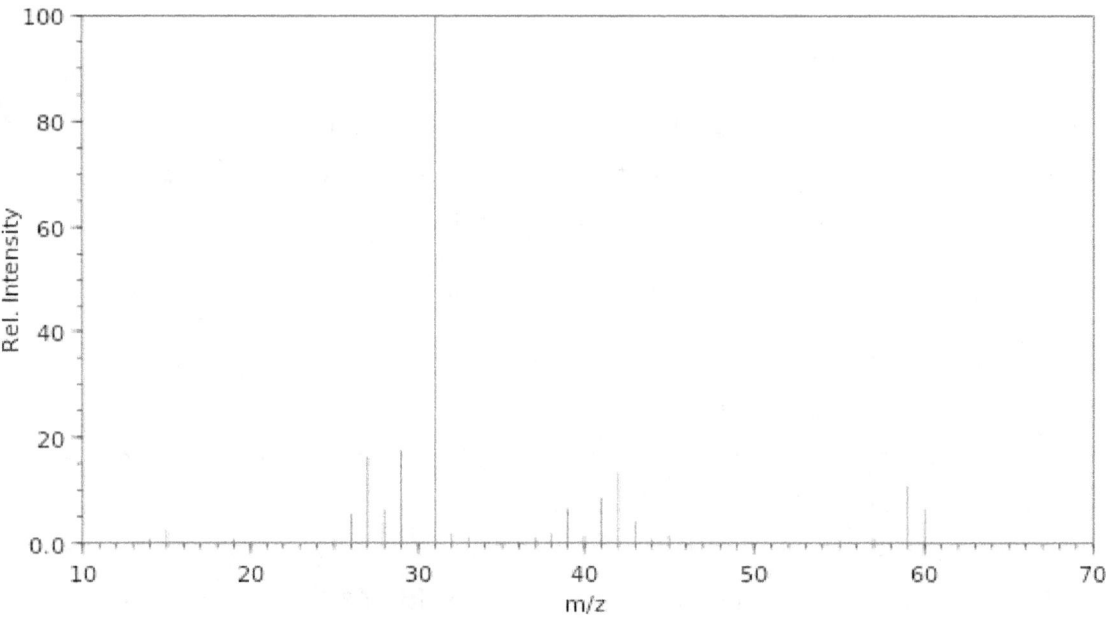

As before, it is important to check if the molecular ion M+ is present and to then consider how the molecular ion could fragment and calculate the mass/charge ratios. Some of the possible fragments are indicated by the dashed lines and the masses of the fragments are shown below the diagram.

1 – propanol	2 – propanol

Fragments:

	1-propanol		2-propanol
	- CH_3 (m/z = 15)		- CH_3 (m/z = 15)
	- C_2H_5 (m/z = 29)		- C_2H_5O (m/z = 29)
	- C_2H_5O (m/z = 45)		- C_2H_5O (m/z = 45)
	- C_3H_7O (m/z = 59)		- C_3H_7O (m/z = 59)

as well, in both cases, the molecular ion $[C_3H_8O_5]^+$ (m/z = 60).

The mass spectrum of the molecule we are considering is shown below. All the fragments are present and it is extremely difficult to differentiate between the two possible structures. This leads us on to the theory and application of Nuclear Magnetic Resonance (nmr) spectroscopy, which is discussed next and is used to answer this question.

84

Chapter VII

Nuclear magnetic resonance spectroscopy

Nuclear magnetic resonance spectroscopy is by far the most powerful technique for probing the structure of an organic molecule but it was not discovered by chance. In fact, it was designed by physicists on the basis of the structure of the atom and the first spectrometer was built in the 1930s.

Hydrogen NMR (^{1}H NMR) allows us to determine the precise arrangement and number of hydrogen atoms on specific carbon atoms whilst *Carbon (^{13}C) NMR* spectroscopy identifies specific type of carbon atoms. Together, these analyses allow us to determine the precise and exact arrangement of atoms within a molecule.

To understand the fundamental principle it is necessary to return to the concept of the atom before we consider the history of the development of this most important technique and its applications. This requires us to appreciate the concept of *nuclear spin* which involves the resolution of the question of why an orbital can only contain a maximum of two electrons even though both negatively charged particles repel each other. It also explains why an orbital can only contain two electrons.

Consequently, this chapter commences with a discussion of *nuclear spin* before the concepts and application of *^{1}H NMR spectroscopy* and *^{13}C NMR spectroscopy* are discussed.

Nuclear Spin

To explain nuclear and electronic *spin* we need to revisit the earliest discoveries of electricity and magnetism.

Hans Christian Ørsted (pictured left) was a Danish scientist who discovered that electric currents create magnetic fields. He found that if two wires carry an electric current, which we now know to be the movement of electrons, in the same direction then they attract each other and the wires are pulled slightly towards each other. On the other hand if the wires contain electric currents travelling in opposite directions the wires repel each other and move slightly apart from each other. Ørsted demonstrated that this attraction and repulsion was due to magnetism by placing a small compass in multiple positions adjacent to the current – carrying wires and recording the deflections of the needle. He thus became the first person to connect electricity and magnetism and this led to the concept of electric and magnetic fields which, in turn, led to the determination, mathematically by James Clerk Maxwell, that electromagnetic radiation was created by a combination of the two fields. The key point for us is that a moving charge generates a magnetic field.

The converse must also also be true i.e. a magnetic field can make a charged particle move simply because there is no reason for that not to occur. In this context movement does not necessarily mean physical movement from one position to another. It can simply mean the particle is rotating on its axis. Ørsted's discovery revealed a fundamental connection between charged particles which can, mathematically be considered as an electrical field and a magnetic field. Both the electron and the nucleus are moving charges since they rotate i.e. spin. As with a child's spinning top, the electron can spin in a clockwise or an anti-clockwise direction. The spinning electron charge generates a magnetic field since it is moving i.e. rotating. It does not have to change physical location to be moving. If one electron spins in a clockwise direction it will generate a magnetic field in an upwards direction. Likewise, an electron spinning in an anti-clockwise direction will generate a magnetic field in the opposite direction. The two opposing magnetic fields will repel each other but the repulsion is overcome by the attractive power of the nucleus so they are forced to remain in the orbital.

This principle though does explain why no orbital can contain more than two electrons.

In a filled orbital i.e. one containing two electrons, the electrons spinning in opposite directions will generate opposing magnetic fields but be kept in place. On the other hand, a third electron, also spinning, will generate a magnetic field in one direction or another and will be repelled by one of the two electrons. This explains why an orbital can only contain a maximum of two electrons.

Just as with electrons, a spinning nucleus generates a magnetic field. If we consider the ^1H nucleus, it is simply a proton and can spin in a clockwise or an anti-clockwise direction. In the absence of an applied, external magnetic field, on average, 50% of the hydrogen nuclei will spin in one direction and 50% in the opposite direction.

Under certain circumstances the spinning nucleus can be made to change its spin from clockwise to anti-clockwise and vice versa and is explained later as this principle forms the basis of nuclear magnetic resonance and is discussed below.

If we picture the atom to comprise a small dense, positively charged nucleus surrounded by electrons then we can envisage the nucleus rotating in a clockwise direction or an anti-clockwise direction. The electrons will also be spinning on their axes but their analysis is an entirely different discipline, *Electron Spin Resonance Spectroscopy*, which is not relevant for the purposes of these volumes.

We know that a moving charge generates a magnetic field and rotation of a positive charge is as much movement as is physical translation from one physical location to another.

The magnetic fields generated will point in opposite directions. Since the nucleus rotates at a slight angle the magnetic field described by the rotating nucleus is a circle. This is analogous to the rotation of the Earth where magnetic North is not literally at the top of the planet.

This process is known as *precessing* and is best considered visually as demonstrated on the right.

For mathematical reasons, the spin of the nucleus is denoted as either $+\frac{1}{2}$ or $-\frac{1}{2}$, abbreviated to $\pm\frac{1}{2}$. Given that the magnetic field can point up or down the rotation of the nucleus is described as *spinning up* or *spinning down*. For even greater simplicity, the entire atom can be described symbolically simply as an up arrow or a down arrow. In the absence of a strong magnetic field (that of the Earth is too weak to be of any effect) the atoms are all of a jumble and will precess in all 360° directions and can be represented in the following manner, with each arrow indicating the direction of the magnetic field of an atom.

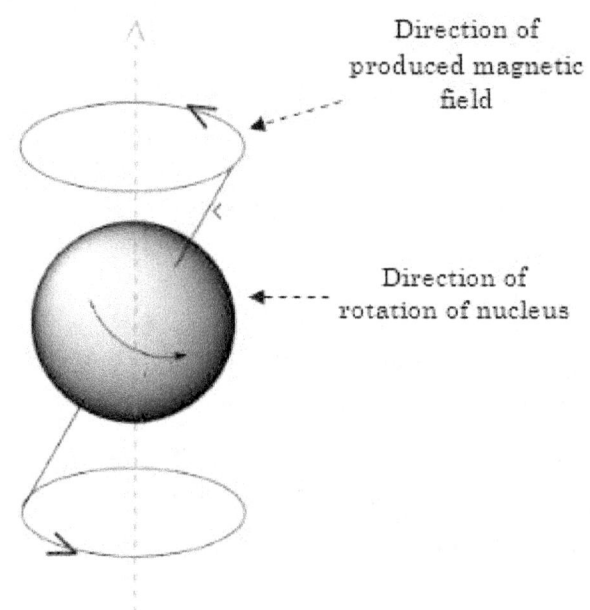

Direction of produced magnetic field

Direction of rotation of nucleus

In the presence of a strong magnetic field, however, all the atoms can be made in orient in the same direction and, where B is the strength of the applied magnetic field, can be represented as follows:

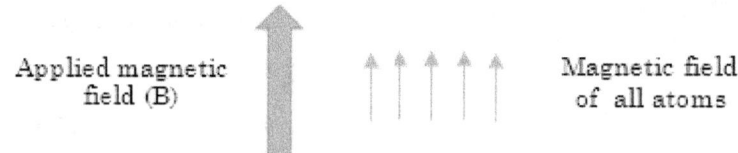

Applied magnetic field (B)

Magnetic field of all atoms

This is the lower, $+\frac{1}{2}$, energy state of the atoms and it was theorised that if the aligned atoms were irradiated with radiowave frequencies (rf) the atoms could all be made to flip from the lower energy ($+\frac{1}{2}$) orientation to the, higher, $-\frac{1}{2}$, spin state:

This transition can also be represented as shown below:

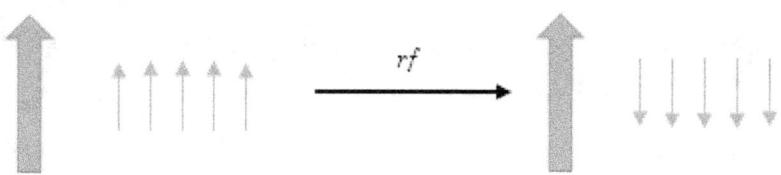

rf

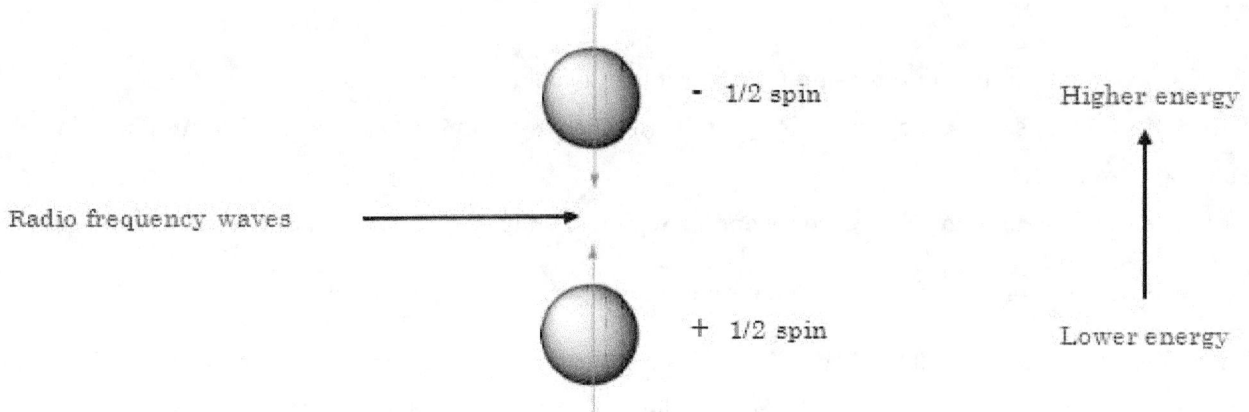

To achieve this the magnet has to be extremely powerful, many hundreds of thousands of times more powerful than the Earth's magnetic field, but a fortunate consequence of this is that the frequency of the radiowave radiation required to flip the nucleus is in the radiowave region of the electromagnetic spectrum, specifically UHF (TV) and VHF (radio).

This idea was first conceived, in 1936, by the Dutch physicist Cornelis Gorter but he could not secure the finances necessary to build the required equipment. His concept was demonstrated one year later by Isidor Rabi (Nobel Laureate, 1944). Rabi established the principle by demonstrating that a stream of lithium ions could be flipped from the lower to the higher energy state.

The next significant development was made by Felix Bloch and Edward Purcell who independently, and almost simultaneously, demonstrated the phenomenon using paraffin wax and water respectively. They were the joint Nobel Physics Laureates for this discovery in 1952.

The huge number of Nobel Prizes awarded for this research demonstrates that the potential of nuclear magnetic resonance has never been underestimated.

This was swiftly followed by the discovery that different hydrogen atoms attached to different carbon atoms, generating specific magnetic fields could be flipped, and be detected and identified individually.

In each circumstance, the specific radio frequency is unique to that specific environment and meant that each individual chemical and magnetic environment could be differentiated by measuring the specific frequency at which these flips occur. A further conclusion was that the measurements also indicated the relative number of hydrogens of each type.

A question often asked is why certain nuclei demonstrate this phenomenon and others do not. For example, the ^{1}H isotope does resonate with the magnetic field but the ^{2}H isotope does not. With regard to carbon, ^{12}C does not but ^{13}C does.

The reason for this is straightforward.

The *nucleons*, neutrons and protons, all spin within the spinning nucleus. If the number of nucleons is an even numbered integer then the spins of these nucleons cancel each other out and the nucleus does not experience any effect of the magnetic field.

For example,

- The ^{12}C isotope contains 6 protons and 6 neutrons and so does not resonate. The ^{13}C isotope has the same number of protons (6) but 7 neutrons and so has an odd number of nucleons and does resonate.

- ^{1}H has one proton and no neutron so resonates whilst ^{2}H has one proton and one neutron and does not.

The lack of resonance of the ^{2}H and ^{12}C has both advantages and disadvantages:

- The ^{1}H isotope comprises 99.98% of all hydrogen atoms whilst ^{2}H comprises a mere 0.2%. This means that the ^{1}H nmr measurement provides a very clear, detailed and strong spectrum in a matter of minutes.

  Although analysis of solid samples is possible, for organic compounds, all nmr measurements are conducted on a sample dissolved in a suitable solvent.

  Most organic compounds dissolve in either water or a solvent such as chloroform. If dissolved in ordinary water or chloroform then nothing would be detected in the nmr experiment since the ^{1}H isotopes present would completely swamp the signal from the sample since the solvent molecules are present in many trillions of times in excess of the sample molecule.

  That the ^{2}H isotope does not resonate means that, if the compound is water – soluble then it can be dissolved in $^{2}H_2O$, also known as heavy water (D_2O), whilst those soluble in organic solvents can be dissolved in deuterated chloroform $C^{2}HCl_3$, also and less clumsily written as $CDCl_3$.

  Note that, traditionally, many chemists have referred to ^{2}H as deuterium but the use of specific names and symbols for ^{2}H (deuterium) and ^{3}H (tritium) is now discouraged since there is no rationale for the three isotopes of hydrogen to have different names whilst no isotopes of any other element do. There are, however, times when it is more convenient to use those symbols and chemists use whichever is simplest even if it is inconsistent with stated principles hence the common use of $CDCl_3$ and D_2O.

- The most common carbon isotope (^{12}C) which comprises approximately 99% of all carbon atoms does not contribute to the nmr spectrum whilst ^{13}C does. Since ^{13}C comprises only 1% of all carbon atoms, the ^{13}C nmr spectrum takes many hours to collect and is still very noisy.

The lack of resonance of the ^{12}C isotope does, however, confer its own advantage.

Typically, the spectral range is from 0 – 220 ppm (discussed below) and the distinct chemical and magnetic environments enable rapid differentiation of, and distinguishing between, different environments.

Moreover the tiny proportion of ^{13}C isotopes means that it is highly unlikely that two such isotopes will be present in the same molecule to any great extent and this means that their magnetic fields do not interfere with each other and so there are no doublets, triplets etc; as discussed below.

Chemical Shift

A note about the units of the spectra.

The strength of the magnet is fixed and, consequently, all measurements are of the frequencies of the radio waves that are absorbed. NMR spectrometers operate at different frequencies. Those producing low resolution spectra typically operate at 90MHz with relatively low strength magnets whilst high resolution machines operate at 200MHz, 300MHz, 360MHz, 400MHz and 500MHz. The magnets involved are huge in order to produce the fields needed and however high the quality of the manufacturing it is impossible to ensure that they are all consistently manufactured. The slightest difference in the magnet can vary the magnetic field sufficient to produce shifts at different frequencies.

The solution is to measure the frequencies relative to a known standard material.

The standard used is tetramethylsilane (TMS) which has the formula $Si(CH_3)_4$. This produces a very strong signal due to the presence of 12 hydrogen atoms per molecule and since the molecule is tetrahedral all the hydrogen atoms are both *chemically* and *magnetically* equivalent so produce a sharp singlet.

$$
\begin{array}{c}
CH_3 \\
| \\
H_3C - Si \cdots CH_3 \\
| \\
CH_3
\end{array}
$$

The molecule is tetrahedral and, in a two dimensional diagram, the hatched line represents a group behind the page whilst the broad arrow points in from the front of the page.

All measurements are converted into the *chemical shift* (δ) using the formula for each peak:

$$\text{Chemical shift } \delta = \frac{\text{frequency of signal} - \text{frequency of TMS}}{\text{Spectrometer frequency}} \times 10^6$$

Since, in this case the frequency of the signal is equal to the frequency of TMS, the numerator is zero which means that whatever the frequency or the strength of the magnet, TMS always appears as zero and all other measurements are calculated relative to the $\delta 0$ ppm (parts per million).

The multiplication by 10^6 is simply to ensure the numbers are manageable since the frequencies are in the Megahertz (MHz) range.

The use of the units, whilst correct, is often neglected again for simplicity with the assumption that all those analysing the spectra understand that the chemical shift does have units.

The *chemical shift* has a number of advantages.

Firstly it takes account of manufacturing differences, as explained before, but for a chemist interpreting a spectrum the range of δ values, $0 - 12$ ppm for ^{1}H spectra and $0 - 220$ ppm for ^{13}C spectra, is much easier than considering huge frequencies which few people can grasp, appreciate and compare.

It should also be noted that, although ^{1}H and ^{13}C measurements are by far the most commonly performed nmr spectroscopic measurements, other nuclei including ^{10}B, ^{11}B, ^{14}N, ^{17}O, ^{19}F and ^{31}P do resonate but are much less common and are not considered in the other two volumes of this series.

The most effective way to learn to interpret spectra is to analyse some and this is what we will now do, starting with ^{1}H and then continuing with ^{13}C nmr spectra.

^{1}H NMR spectroscopy

Chemical and magnetic equivalence

We will consider *chloroethane* (CH_3CH_2Cl)

which can only exist in the one form since the structures displayed below are identical as one can be turned over to appear the same as the other i.e. they are mirror images and are identical.

It is, however, clear that there are two different types of hydrogen atoms as indicated below:

Given that these hydrogen atom groups experience different magnetic fields they can be distinguished since they will resonate at different frequencies.

1. The ^{1}H nmr spectrum will therefore, as a first step, comprise two peaks due to the:-

    ▪   – CH_3 group

    ▪   – CH_2 group

    The position of these peaks are stated as chemical shifts which is a concept discussed below.

2. Secondly since the intensity of the resonant signal is proportional to the number of hydrogen atoms, the areas under the peak will be in the ratio of 3:2. The measurements of the areas (known as the *integrals*) are not absolute and are relative to each other.

This means that we can now expect the spectrum to comprise two peaks with one of 50% area greater than the other. This immediately indicates which peak is due to the – CH_3 group (the larger peak by area) and which is caused by the – CH_2 group (the smaller peak again by area).

However there is one further great benefit to the ^{1}H nmr experiment and that is due to interactions between the magnetic fields of hydrogen atoms on adjacent carbon atoms, the *n+1 rule*.

Peak Multiplicity

The **n+1 rule** states that a single peak will be split by the number of hydrogen atoms on adjacent hydrogens. This is due to the atoms either spinning up or spinning down.

If we consider a single hydrogen atom then if there were no adjacent hydrogens the hydrogen will produce a single peak. However, if there is one hydrogen atom on one of the immediately adjacent carbon atoms, that hydrogen atom may itself be spinning up or down. These create slightly different magnetic environments. Within the entirety of the sample then, on average, there will be equal number of hydrogens spinning up and spinning down. Although *chemically equivalent*, these hydrogen atoms are **not** *magnetically equivalent* and will resonate at slightly different frequencies.

Consequently a lonely hydrogen atom's single peak, a *singlet*, is split into two peaks of equal height which is known as a *doublet*.

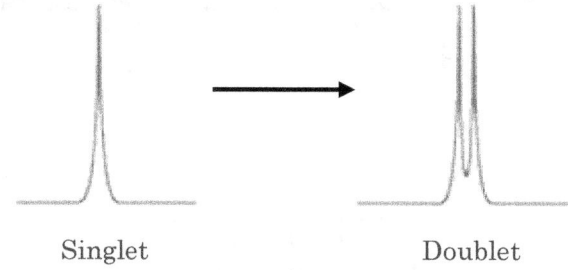

Singlet   Doublet

The area under the doublet, the *integral*, is equal to the area under the singlet.

The same principle applies to a hydrogen atom (highlighted) with two hydrogen atoms on adjacent carbons. It makes no difference if the hydrogens are all on one of the immediately adjacent carbon atoms or are distributed across both i.e. both scenarios produce a triplet since both there are two hydrogens on adjacent carbon atoms.

$$H - C \begin{array}{|c|} \hline C \\ | \\ H \\ \hline \end{array} C - H \qquad - C \begin{array}{|c|} \hline C \\ | \\ H \\ \hline \end{array} \begin{array}{c} C - H \\ | \\ H \end{array}$$

This means that both of these situations will produce a triplet.

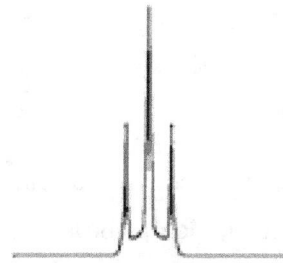

as, in both cases, the analysed hydrogen has two hydrogen atoms on immediately adjacent carbon atoms.

As above the area under the triplet equals that of the area of what would have been a singlet.

The process continues with a singlet split into a:

- *quartet* when there are **three** hydrogens on the immediately adjacent carbon atoms:

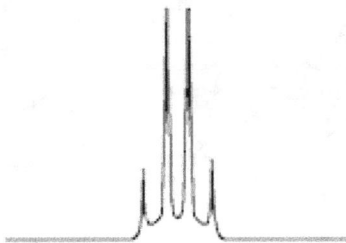

- *quintet* when there are **four** hydrogens on the immediately adjacent carbon atoms:

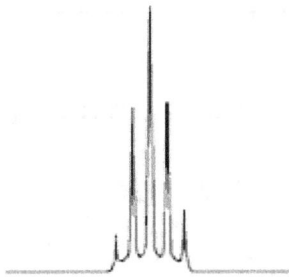

The doublets, triplets, quartets etc; are collectively referred to as *multiplets*.

In sum, the number of hydrogens splits the singlet into a peak with the *number of adjacent hydrogens +1*.

This is the n+1 rule.

This rule immediately identifies how many hydrogen atoms are on immediately adjacent carbon atoms which also demonstrates that the effects do not exist beyond two bonds.

If we reconsider the highlighted structure of the molecule,

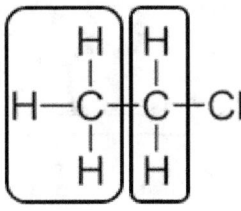

we may now conclude that there are:-

- **Two** types of chemically and magnetically equivalent hydrogen atoms and so **two groups of peaks**.
- If we consider the $-CH_3$ group, then the peak of relative area **3**, must be a *triplet* as there are two hydrogen atoms on the adjacent carbon.
- With regard to the $-CH_2$ group, then the peak of relative area **2**, must be a *quartet* as there are three hydrogen atoms on the adjacent carbon.

Essentially, therefore, the ¹H nmr spectrum will comprise these two groups. When we perform the actual measurement we obtain the following spectrum:

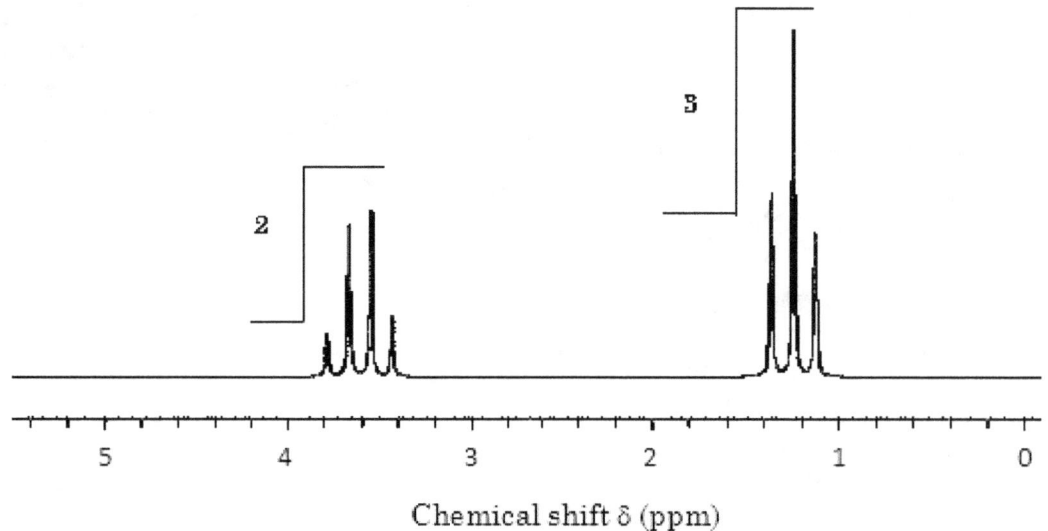

Chemical shift δ (ppm)

It is, therefore, perfectly clear which multiplet is produced by each set of C – H atoms.

The spectrum of chloroethane demonstrates the power of the ¹H nmr spectroscopic technique. The other real power is that the different types of hydrogen atoms can be identified using a correlation chart since they appear, consistently for a huge range of compounds in the same region.

A useful example is shown below:-

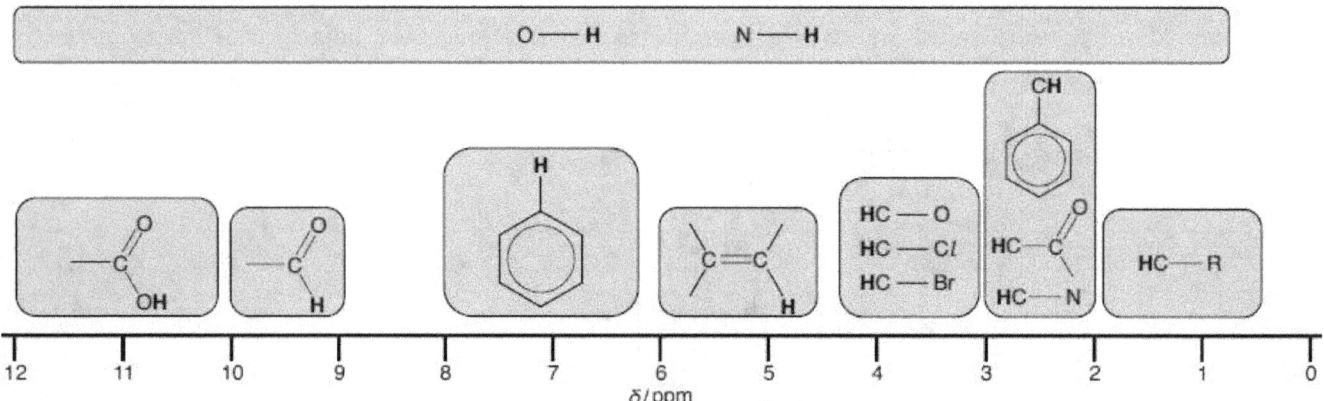

The only difficulty arises from the labile O – H and N – H hydrogen atoms since they can appear almost anywhere in the spectrum.

There are other matters to note about the spectra which are not necessary for the interpretation of spectra in the following two volumes of this series but are extremely interesting.

Peak heights in a multiplet

If we consider the patterns and shapes of a multiplet, from a singlet to a septet,

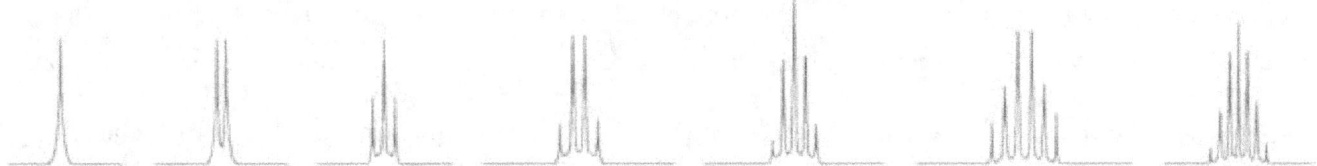

It becomes clear that within the multiplet there is a definite and clear pattern in the peaks:

Multiplet	Number of peaks	Ratio of peak heights
Singlet	1	1
Doublet	2	1:1
Triplet	3	1:2:1
Quartet	4	1:3:3:1
Quintet	5	1:4:6:4:1
Sextet	6	1:5:10:10:5:1
Septet	7	1:6:15:20:15:6:1

The ratio of peak heights is the pattern of integers famous in the Western world as Pascal's Triangle. This pattern has been known for at least 1,000 years though; in Iran it is known as the Khayyam Triangle whilst in China it is called Yang Hui's Triangle.

Blaise Pascal (1623 – 1662) became famous for his book describing a huge number of applications in business, pure mathematics and probability and his name became attached to the triangle even though it had been known for hundreds of years. Pascal was also renowned for his work on vacuums and on pressure and the unit of pressure, mathematically Newtons per square metre, N/m^2, is termed the Pascal (Pa) in his honour. He is also credited with being one the first inventors of the mechanical calculator.

Even though Pascal cannot possibly have considered the application in a technique of which he had no conception it is extremely useful in predicting the multiplet peaks in complex molecules. When we have a molecule with a number of distinct but similar groups the peaks, although being chemically and magnetically equivalent, will appear in the same region of the ^1H nmr spectrum and often overlap. Knowing the ratios of the peak heights allow us to separate out the multiplets and assign them appropriately.

For example, all alkane hydrogen atoms appear between δ 0.5 and δ 1.9 ppm. With a large molecule there will be a significant number of peaks in that region and the multiplet peak heights enables them to be distinguished.

There is another aspect to the multiplets and that is the distance between the peaks in a multiplet.

This is due to a phenomenon known as the *coupling constant* which is discussed next.

Coupling Constants

When protons communicate with each other the effect exerted by one on another must be reciprocated. This is measured by calculating the distance between the peaks of a multiplet. Two groups of protons on nearby carbon atoms will have the same *coupling constant*, referred to as J, and which is measured in Hertz (Hz).

If they do not have the same coupling constant then they are not communicating and hence **not** adjacent.

If we consider 1,3 – dibromopropane

Since the molecule is symmetrical around the central – CH_2 group, We can immediately see that there are two sets of chemical and magnetically equivalent groups:

These are identical to each other but different to the central – CH_2 group which is circled:

The structural formula and the molecular formula $BrCH_2CH_2CH_2Br$ enables us to predict that there will be two multiplets in the ^1H nmr spectrum in the ratio 2:1.

- One group will be due to the hydrogens on the terminal carbons of which there are four.
- The second group will be due to the central, circled, hydrogens of which there are two.

Since we use the smallest whole number ratio the areas under the groups will be in the ratio 2:1

Referring to the correlation table, we can see that the hydrogens on the terminal carbons will appear between δ 3 ppm and δ 4.3 ppm whilst the central – CH_2 group appears between δ 0.5 and δ 1.9 ppm

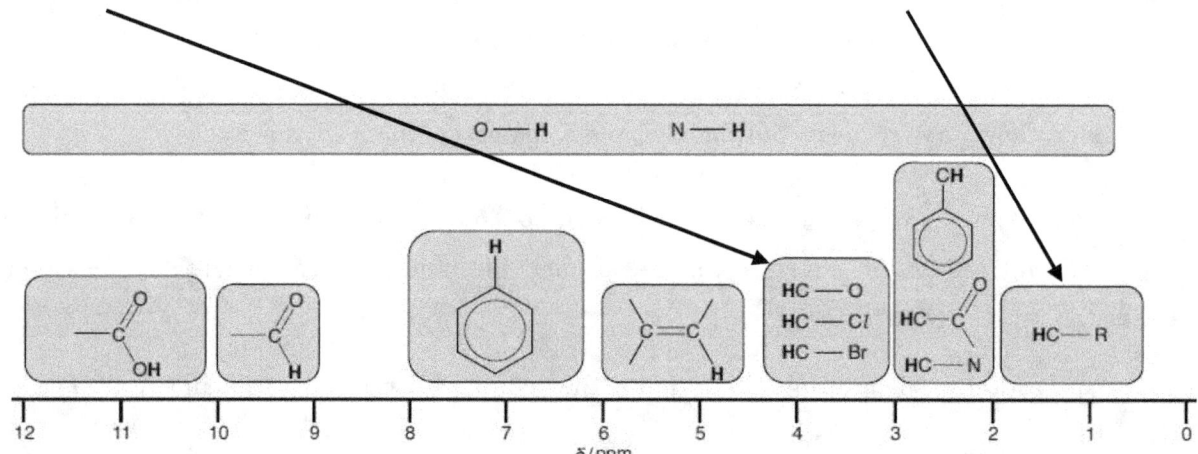

As well as this, with our knowledge of the *n+1 rule*, we can, also predict that since

※ the central –CH$_2$ group has a total of four hydrogen atoms on adjacent carbon atoms, this will produce a quintet and with our knowledge of chemical shifts, this will appear between δ 3 and δ4 ppm.

※ The terminal – CH$_2$(Br) peak will generate a triplet since both ends of the molecule are identical and each of the hydrogen atoms on the group have two hydrogens on the central carbon atom and will be somewhere around δ 2 ppm. Due to the presence of the electronegative bromine atoms, it will be deshielded and will appear to the extreme left of the alkyl region of the correlation chart.

The actual ^{1}H nmr spectrum is as shown below:

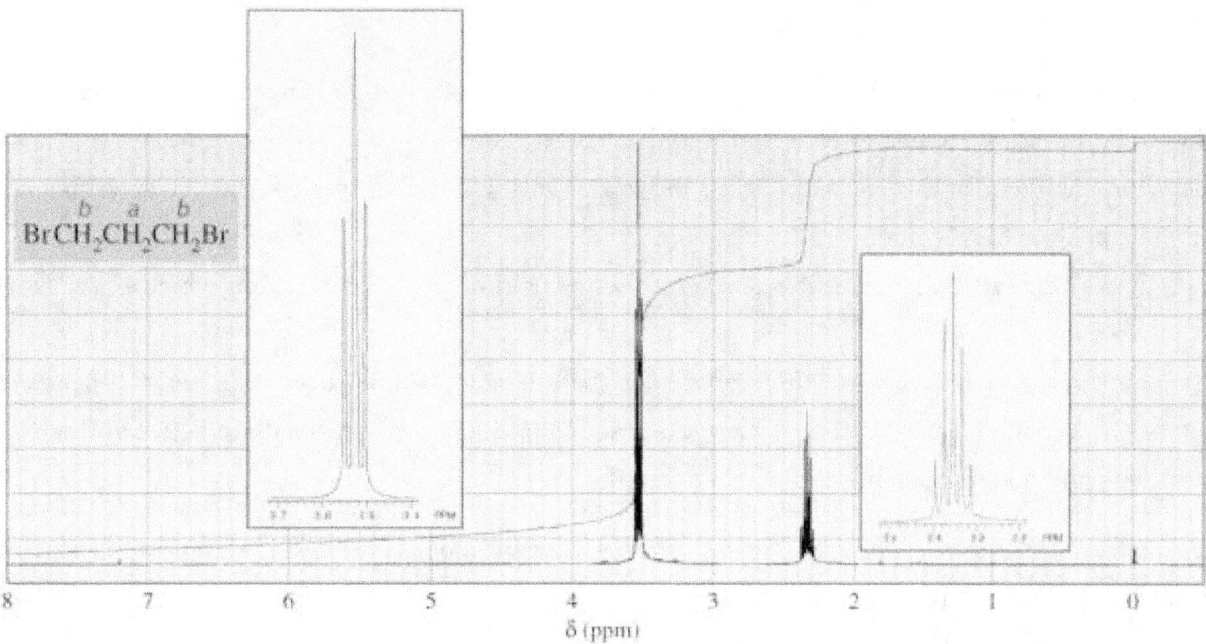

The multiplet peak shapes are as predicted but the – C*H*$_2$Br group is slightly to the left of the region identified in the correlation chart which shows both its use and its limitations. The one used in this volume is suitable for the compounds analysed in Volumes II and III of this series and has been selected for its simplicity. More detailed correlation charts would have predicted this shift but can be overly complicated at one and the same time. No correlation chart can cover all scenarios unless we want a very busy chart which would be difficult to navigate.

What is really interesting from our perspective though is the coupling constants of the two multiplets. They are significantly different, indicating that the protons on the terminal carbons do not communicate with those on the central – CH$_2$ – hydrogens.

Without going into huge detail, analysis of numerous compounds have produced a number of general rules and two examples stand out.

One group that has not been addressed yet are aromatic rings such as benzene and substituted benzene molecules.

The second group comprises the cis – and trans – isomers of alkenes.

Both of these groups inform us of their structures through the chemical shift and the coupling constant.

The ^{1}H nmr spectroscopy of benzene

Benzene is famous for its stability which can drawn as two, Kekulé, structures:-

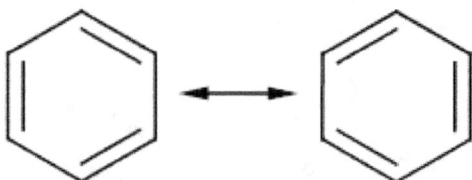

where the double headed arrow represents two possible and equivalent structures

This, however, cannot be correct for four reasons:

- The benzene molecule would react with three mole equivalents of bromine water which it does not
- There should be two types of chemically and magnetically equivalent hydrogen atoms so the ^{1}H nmr spectrum should show two peaks of equal integral. This is not what occurs and the spectrum contains only one peak at δ7.35 ppm.
- Equally significantly, this peak is significantly downfield of the alkene region which is typically from δ4.5 – 6.5 ppm
- From our knowledge of bond lengths, would have three long C – C and three shorter C = C bonds. In fact all the C – C bond lengths in the ring are the same length and are intermediate between the lengths of the single and double carbon – carbon bond lengths.

The reason for this is the overlapping of the p – orbitals creating a delocalised orbital containing six electrons, one from each of the constituent carbon atoms.

The six p – orbitals overlap (below left) creating a ring above and beneath the planar carbon skeleton in which the six p – electrons travel around the ring (below right)

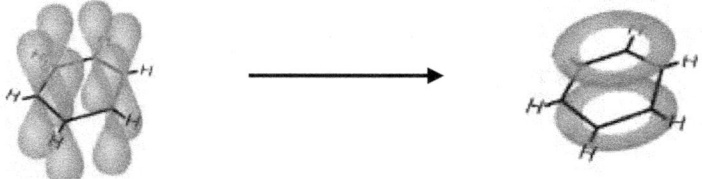

This leads to all the carbon – carbon being equivalent and explains why they are all the same length in what become so called *aromatic* compounds. This also means that, in benzene, all the hydrogen atoms are chemically and magnetically equivalent and the benzene ring is drawn as:

All so called aromatic compounds resonate in the region δ7 – δ8 and this means that any peak present there can be immediately be assigned to the presence of an aromatic ring. This is important when we consider the structure of benzaldehyde, the natural almond flavouring, in Volume III of this series.

The other interesting situation, from our perspective is the isomerism of alkenes which we consider next.

The ^{1}H nmr spectroscopy of isomers

We can take bromochloroethene as an example. Since the C = C bond is rigid, due to overlap of the p – orbitals, there are two possible structures:

cis   trans

Cis and trans are old terminologies (*cis* = together i.e. on the same side whilst ***trans*** = across i.e. on opposite sides of the C = C which have been pretty much replaced by (E) – and (Z) – terms but they are still used.

Using our knowledge of spectroscopy and spectrometry, we can immediately determine that:-

* The infrared spectrum would identify the presence of the C = C and C – C bonds (in the fingerprint region) with, of course, the useless C – H stretches present in all organic molecules.
  In addition, there will be C – Cl and C – Br peaks in both spectra so that doesn't help.
* The mass spectrum would produce the same fragmentation patterns for both so that doesn't help.
* Superficially, the ^{1}H nmr spectra would appear the same with the same peaks with the same chemical shifts since each hydrogen is attached to a carbon atom bonded to a halogen but here comes the difference: the coupling constants in the spectra are found to be different.

Naturally the ^{1}H nmr spectra of the two isomers will be similar but it is found that the coupling constants of cis – and trans – isomers are different. It's a small difference but is nevertheless highly significant. The coupling constant, **J**, (the distance between the peaks of a multiplet) is smaller for cis – isomers than for trans – isomers and typically the differences are as shown below:

Isomer	J (Hz)
Cis	6 – 15
Trans	11 – 18

There is clearly some overlap but it is clear that any J value below 11Hz indicates that the molecule possesses a cis – structure whilst anything above 15Hz demonstrates that the isomer must be trans. Reverting to aromatic compounds, we can consider the three possible isomers of substituted benzenes. Nowadays we number the arrangements but they are traditionally known as ortho – , meta – and para – as shown below:

Coupling constants: Ortho: 7 – 10 Hz   Meta: 2 – 3 Hz   Para: 0 – 1 Hz

It is clear that we can use the coupling constant, J, to distinguish between the isomers of a substituted aromatic ring.

To sum up then, with ^1H nmr spectra we can identify the position and numbers of individual hydrogen atoms in a molecule by way of:-

- Chemical shift δ ppm;
- Area under the peak, the integral;
- Multiplicity of the signal;
- The coupling constant, **J**.

That may appear to be everything we can learn about the structure of a molecule but that is before we consider the application of ^{13}C nmr spectroscopy which is discussed next and is thankfully much simpler than ^1H nmr spectroscopy!

¹³C NMR spectroscopy

The interpretation of a ¹³C nmr spectrum is, in many ways, simpler than the ¹H spectrum for the simple reason that the predominant ¹²C isotope (99% of all carbon atoms) does not resonate with the electromagnetic spectrum and so the measurement is of the ¹³C isotope only. The reason for this is explained above.

As also discussed above, given its relatively rare occurrence, it is highly unlikely that two such isotopes will exist in the same molecule so there will be no couplings between the atoms and the peaks in a ¹³C nmr spectrum are only ever singlets. That might appear a disadvantage but this is overcome by the range of chemical shifts in such a spectrum as shown below:

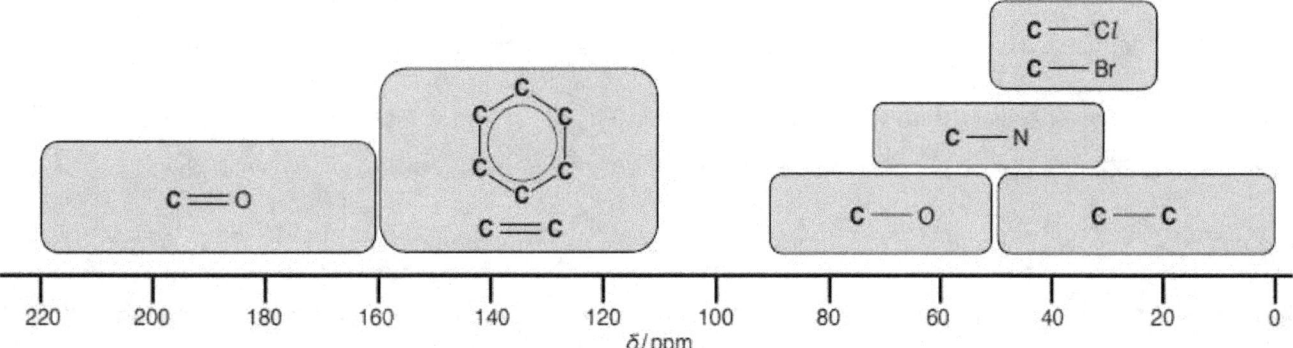

There is therefore rarely much difficulty in assigning the peaks to specific carbon atoms in small molecules.

If we refer to chloroethane again we can predict that there will be two peaks of equal area and this is indeed what we observe:

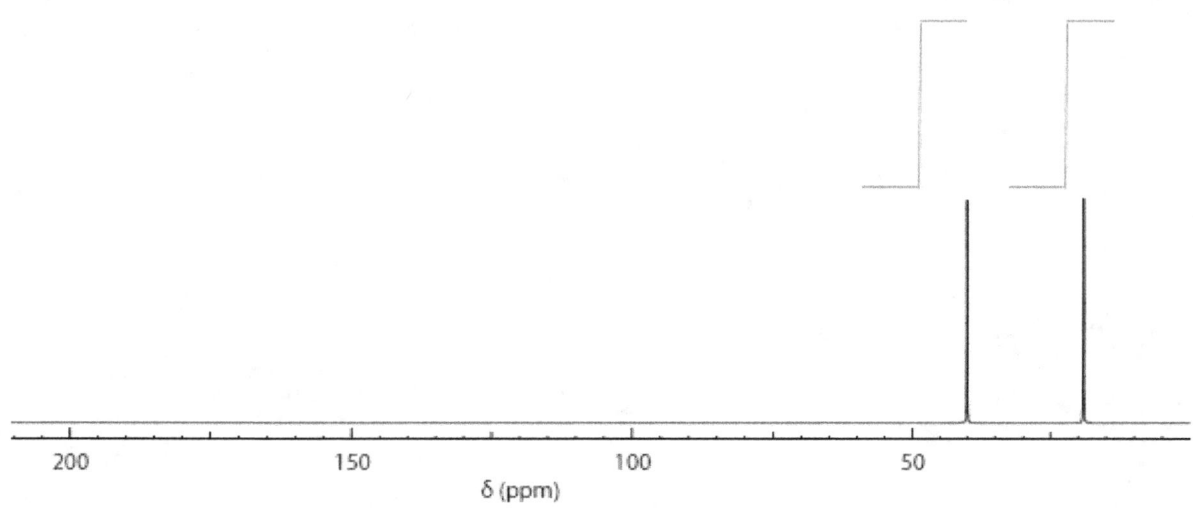

From the correlation chart we can conclude that the peak at δ40 ppm is due to the chlorine substituted carbon, – CH_2Cl, whilst the peak at δ19 ppm is due to the alkane carbon, – CH_3.

Although the regions overlap the C – Cl carbon does not appear below δ20 ppm so it is straightforward to distinguish between the groups.

There is no coupling due to the unlikelihood of two ¹³C isotopes co-existing adjacently but the area under the peak, the integral, applies just as it does in the the ¹H nmr spectra.

The ^{13}C nmr spectroscopy of benzene

If we reconsider the Kekulé structures of benzene,

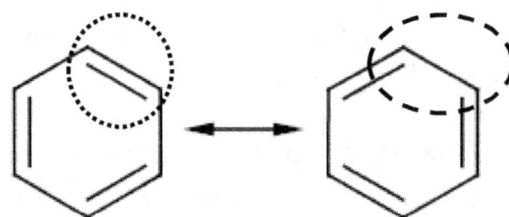

we can can see that there are two types of carbon atom,

- three that are doubly bonded C = C, circled with a dotted line (above left) and
- three which are singly bonded, C – C, identified by the oval with a dashed arrow (above right).

If the above structure is correct then we should observe two peaks of equal integral, one in the region δ 115 – 160 ppm, assignable to a C=C bond, and one in the region 0 – 50 ppm due to the presence of the C – C bond.

That is not what we observe and the ^{13}C nmr spectrum comprises one peak, a singlet, at δ 144.5 ppm.

We have now explored the principles, application and power of spectroscopic and spectrometric analytical techniques and practical examples are exemplified, for small molecules, in Volume II and natural flavourings in Volume III. In both handbooks, the first few examples are worked through in full with the remainder to practice on, with full answers at the end of each handbook.

The following chapter records, as pen portraits, the lives and achievements of the scientists and engineers who completely changed our view of the world.

Chapter VIII

Pen portraits

This chapter describes the people who played a major part in developing our concepts of the atom from the conception of a hard, indivisible solid particle to learning its internal structure and which led to a century of chemical discoveries.

Prestigious in their own right for their discoveries, it is intriguing to learn about their other achievements in flying, sport, music, art, mountaineering and the development of nuclear weapons, an undesired consequences of the discoveries.

Anders Jonas Ångström

Medelpad - born **Anders Jonas Ångström** (1814 – 1874) is regarded as one of the founders of the science of spectroscopy. Educated at Uppsala University, he was appointed to a lectureship there in 1839 but in 1842 moved to the Stockholm Observatory and in 1843 he was appointed Keeper of the Uppsala Astronomical Observatory.

Although initially working on terrestrial magnetism, his investigations changed to the study of heat conduction and spectroscopy. Ångström determined that an electric spark produces two superposed spectra, one from the metallic and the other from the gas which separates the electrodes. This led him to conclude that the gas emits lights rays of the same refrangibility (refractability) as those it can absorb.

This is analogous to the later observations that emission and absorption spectra are complementary. His work was conducted at approximately the same time as similar studies by Robert Bunsen and Gustav Kirchhoff.

Appointed to a full professorship in 1858, Ångström began to study the solar spectrum and combined he was one of the first to combine spectroscopy with photography spectroscope with photography. He demonstrated that the sun's atmosphere contains hydrogen. He produced an authoritative map of the solar spectrum and recorded the wavelength of over 1000 lines.

Ångström's measurements were slightly incorrect (by an error of 1 in 8000) since the metre rule that he used for calibration was very slightly shorter than the standard metre.

Ångström was the first to study the the spectrum of the aurora borealis and detected the characteristic bright line in the yellow – green region of the visible spectrum which has been named in his honour. He also suggested that his observations and measurements led to a conclusion equivalent to Kirchhoff's law of thermal radiation and that his results also explained the so-called Fraunhofer lines: dark lines in the spectrum of sunlight.

Ångström created a new unit of measurement 0.1 nm (0.1×10^{-9}m) which is now written as 10^{-10}m and is named the Ångström in his honour.

He died from meningitis, in June 1874, shortly before his sixtieth birthday.

Ångström's son, Knut Johan Ångström, was also a physics professor at Uppsala University. Knut studied the the radiation of solar energy, nocturnal emission of heat from the Earth and its absorption by the Earth's atmosphere and devised a method of obtaining a photographic record of the infrared spectrum.

His work on the absorption of heat by the atmosphere is of great significance in the development of the theory of the greenhouse effect.

F.W.Aston

Professor **Francis William Aston** FRS (1877 – 1945) won the 1922 Nobel Prize in Chemistry for his development of the mass spectrograph, his discovery of isotopes and for his enunciation of the whole number rule.

Born in Birmingham, and educated at the Harborne Vicarage School and then Mason College in Worcestershire, he studied Physics and Chemistry at Mason College, an external college of the University of London and constructed a laboratory in his family home where he studied organic chemistry.

In 1898 he started as a research student under Sir Edward Frankland, studying the optical properties of tartaric acid compounds and then researched the science of brewing at W. Butler & Co. Brewery in 1900 until he became a research student at Birmingham in 1903. At Birmingham he switched to physics and began investigating radioactivity and x-rays. In 1910 he moved to the Cambridge Cavendish Laboratory to work under Sir J.J. Thomson who was continuing his investigation of cathode rays which he had discovered and the anode rays which had earlier been detected by Eugen Goldstein. During the First World War Aston worked at the Royal Aircraft Establishment in Farnborough developing aeroplane coatings.

Aston returned to Cambridge after the Armistice and resumed work on isotopes and developed his first mass spectrograph which he first reported in 1919. He rapidly improved the instrumentation and by the time he had constructed his third model he had refined the electromagnetic focusing sufficiently to detect the isotopes of neon, chlorine and mercury. In all, Aston detected 212 naturally occurring isotopes and he devised the *whole number rule* which states that '*The mass of the oxygen isotope being defined as 16, all the other isotopes have masses that are very nearly whole numbers*. For these huge accomplishments, he was awarded the Nobel Prize for Chemistry in 1922.

Away from his research, Aston was a noted cross-country skier and ice skater, mountaineer and a keen cyclist. He was also a successful tennis player, golfer and swimmer and, in 1909, went to Honolulu to learn to surf. Aston developed his own internal combustion engine and built his own motor car in 1903 in which he competed in motor races.

A skilled photographer, which was essential to his work on mass spectrometry, Aston was also interested in astronomy and made trips to Canada, Sumatra and to Japan to observe and photograph solar eclipses.

Aston was also an accomplished amateur violinist and pianist and performed in many concerts performed by the various Cambridge University orchestras.

Johann Jakob Balmer

Lausen – born **Johann Jakob Balmer** (1825 – 1898) studied mathematics at the University of Karlsruhe and then completed doctoral research in Berlin and in Basel.

Balmer remained in Basel for the rest of life working as a mathematics school teacher and a lecturer at the University of Basel.

Intrigued by the discovery of individual spectral lines of different elements, Balmer became convinced that there must be a link between them and used the measurements of the hydrogen spectrum reported by Ångström to derive the following formula for computing the wavelength:

$$\lambda = h\,\frac{m^2}{m^2 - n^2}$$

where λ is the wavelength in metres, n and m are integers and h is a constant of proportionality. The units of h had to be metres since integers are unit-less and the wavelength was measured in metres.

It is important to remember that, however complicated an equation, the presence of an equality (=) demands that the overall units must be the same on both sides of the equation.

Balmer found that when n = 2 and m = 2, 3, 4, 5 and 6, he could calculate the wavelengths of the hydrogen spectrum and determined that h, now known as the Balmer Constant, has the value 3.6456×10^{-7} m. It is one matter to devise a formula but science progresses when the relationship can be used to make predictions. Balmer did just this with his calculation for m = 7.

$$\lambda = \frac{h\,m^2}{(m^2 - n^2)} = \frac{h\,(7^2)}{(7^2 - 2^2)} = \frac{3.6456 \times 10^{-7} \times 49}{(49 - 4)} = 3.969 \times 10^{-7}\ \text{m}$$

This meant that he predicted that there should also a line at 397 nm (1 nm = 10^{-9} m).

The presence of this line was confirmed by Ångström and other lines with higher values of the integer m were found in the spectral lines of numerous stars.

This discovery was hugely significant since it confirmed that the lines were related, were fundamental to the hydrogen atom and that hydrogen was present in all the white stars that were studied.

If not fundamental then the lines would have appeared at random, unrelated wavelengths and that the stuff of stars was the same as that on Earth.

Balmer's formula was later found to be a special case of the Rydberg formula which is discussed later in Rydberg's pen portrait although an explanation of the significance of the formula had to wait until Niels Bohr proposed his formulation of the structure of the hydrogen atom.

Felix Bloch

Felix Bloch (1905 – 1983) was awarded, with Edward Purcell, the 1952 Nobel Prize in Physics for the invention of nuclear magnetic measurements which paved the way for the invention of nuclear magnetic resonance spectroscopy.

The Zurich – born Bloch initially studied electrical engineering at the Swiss Federal Institute of Technology but, after attending lectures by Erwin Schrödinger and Peter Debye, he switched to physics. Graduating in 1927, Bloch became Werner Heisenberg's first doctoral student. Heisenberg became renowned for his *Uncertainty Principle* which states that it is possible to know the momentum *or* position of an electron in an orbital but is is impossible to know both simultaneously.

Subsequently, Bloch worked with numerous giants in the world of physics including Wolfgang Pauli, Niels Bohr and Enrico Fermi before moving to Leipzig in 1932.

Some of Bloch's early research interests included ferromagnetism, the wave functions of electrons in solids, and the relationship between temperature and conduction but it was James Chadwick's proof of the existence of the neutron that really sparked his interest and after Otto Stern had shown that the neutron possessed a magnetic moment, Bloch determined to find out why this is the case since, until then, the fact that the proton and the electron both possessed a magnetic moment had been attributed to their charge which is lacking in the neutron

Being Jewish, Bloch left Germany immediately after Hitler came to power, first to Zürich and then to Paris before being invited to Stanford in 1934 where he worked with the 37-inch cyclotron to determine the magnetic moment of the neutron. In 1939, Bloch became a naturalized American citizen and became the first Professor of Theoretical Physics at Stanford.

In 1940, Bloch wed Lore Misch, a physicist specialising in x-ray crystallography and who had also emigrated to America from Nazi Germany. The couple later had four children but shortly after their marriage, Bloch took leave of absence from Stanford to work both on the Manhattan and on radar-jamming technologies.

On his return to academia after the Second World War, Bloch returned to his study of magnetic moments and discovered that radio frequencies could control a magnetic field which in turn could used to incite nuclear excitation; in other words to flip the atom from one state to the other.

Bloch also discovered that atoms in a molecule only resonate at specific frequencies and his discoveries led directly to the concept of the chemical shift which has proven to be fundamental to structural analysis by nuclear magnetic resonance spectroscopy.

In 1954, Bloch became the first Director General of CERN before returning again to Stanford until his final retirement back to Switzerland.

Bloch lived until the age of seventy seven by which time he had been named *the father of medical resonance imaging (MRI)* which had arisen solely from his curiosity about the magnetic moment of the neutron and had developed from nmr spectroscopy.

Niels Bohr

Danish – born **Niels Henrik David Bohr** (1885 – 1962) was awarded the Nobel Prize for Physics in 1922. Bohr was also a philosopher and a promoter of scientific research.

He used the Planck concept of the quanta (packages of energy), Rutherford's model of the atom, the existence of the nucleus and the surrounding electrons etc; as discussed earlier in this volume to develop a new model of the structure of the hydrogen atom.

In 1920, Bohr founded the Institute of Theoretical Physics at the University of Copenhagen and worked with scientists who became world famous such as Werner Heisenberg.

In the years preceding the Second World War, Bohr helped refugees from Nazism and after the German occupation of Denmark, he met Heisenberg again who had become the head of the German nuclear weapons project.

In 1943, word reached him, perhaps originating from Heisenberg, that he was to be arrested by the the Germans but he escaped to Sweden and was then flown to Britain where he joined the British nuclear project and formed part of the British contingent to the Manhattan Project.

After the war, Bohr was instrumental in the formation of Cern. Initially he campaigned for Copenhagen to be its base but, conceding that his choice was unlikely to be adopted, he agreed that it should be based in Geneva. In the post war years he also worked for international cooperation on nuclear energy and to eliminate nuclear weapons.

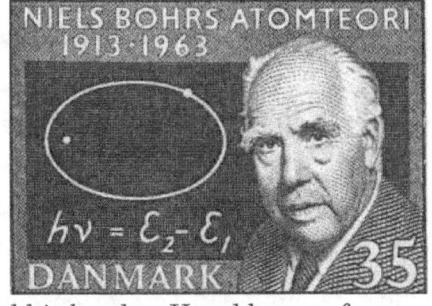

In 1963, Bohr's model of the hydrogen atom, fifty years earlier, was commemorated in two of a series of four Danish postal stamps, one of which is pictured right. the other pair of stamps commemorated Ernest Rutherford.

Niels Bohr came from an extraordinary family. His father, Christian was a famous professor of physiology at the University of Copenhagen whilst his mother, Ellen, was a scion of a wealthy banking and political family and his brother Harald was a famous mathematician and footballer who played for Denmark in the 1908 London Olympics.

Both Niels and Harald played for, the Copenhagen-based, Akademisk Boldklub (Academic Football Club), for whom Niels played in goal.

Niels Bohr won the Nobel Prize in Physics for his theory of the structure of the hydrogen atom. Amongst his other accomplishments, he predicted the existence of a zirconium-like element and *hafnium* was named after the Roman name for Copenhagen.

Niels Bohr is also commemorated by the naming of the element *bohrium* ($_{107}$Bh) after him.

Max Born

Breslau born Max Born (1882 – 1970) was one of the most eminent physicists involved in developing the theory of quantum mechanics and is the one who, with Werner Heisenberg, first established that a solution to the Schrodinger equation resulted in the probability of the electron distribution and, hence, the concept of the orbital.

Born studied Physics at the University of Göttingen, graduating in 1904. Upon graduation he worked on Einstein's theory of Special Relativity and obtained his qualification to become a Professor with, his *habitation* thesis on the Bohr model of the hydrogen atom before being obliged to commence military service but was discharged in 1907 after a severe asthma attack. He then moved to Cambridge where he worked under J.J. Thomson at the Cavendish Laboratory.

By 1914, Born had published twenty seven papers on relativity and crystal lattices and accepted Max Planck's offer of a professorship in theoretical physics at the University of Berlin. He was however subsequently rejected when the physicist first offered the chair, Max von Laue, who had initially rejected it changed his mind and Born moved, instead, to the University of Frankfurt am Main before moving back to the University of Göttingen. However, at the outbreak of the First World War, Born was conscripted as a radio operator in the German Army before being moved to research sound ranging techniques, the sound version of, the later, radar.

In 1921 he returned to Göttingen, which became an extremely significant centre for theoretical physics and it was there that he and Heisenberg formulated the matrix mechanical interpretation of the Schrödinger equation for which he was awarded the 1954 Nobel Prize in Physics. Technically, the solution to the equation produced the probability density distribution of the electrons which are visualised as orbitals. Amongst his research assistants relevant to this series were Friedrich Hund, Werner Heisenberg and Wolfgang Pauli.

In 1918, by chance he met Fritz Haber and together they worked on the concept of energy (enthalpy) changes in a chemical reaction and this is now a standard calculation, the Born - Haber Cycle, which explains the formation and stability of some compounds but, perhaps more importantly, explains why some compounds whose formula can be written down cannot exist. Following the Born - Heisenberg interpretation of the solution to the Schrödinger equation for the hydrogen atom, Einstein wrote to Born and included the now famous remark regarding quantum mechanics: *"Quantum mechanics is certainly imposing. But an inner voice tells me that it is not yet the real thing. The theory says a lot, but does not really bring us any closer to the secret of the 'old one'. I, at any rate, am convinced that He is not playing at dice."* which is often paraphrased as *'God does not play dice'*.

Being Jewish, Born lost his chair at the University of Göttingen when Hitler rose to power and escaped, with his family, to Britain where he took a lectureship at St John's College, Cambridge before moving to become the Tait Professor of Natural Philosophy at the University of Edinburgh in 1936. His daughter, Irene, married an MI5 officer, Brinley (Bryn) Newton-John, who had been born above one of the most famous Cardiff pubs, the Market Tavern, which his mother had run. Following the Second World War the family moved to Australia. Max Born is, therefore, the grandfather of the star of the 1970s iconic film, Grease, and he is also, through marriage, also a great uncle of the comic writer, Ben Elton.

Frederick Brackett

Californian – born **Frederick Sumner Brackett** (1896 – 1988) studied physics at Pomona College and subsequently worked as an observer at the Mount Wilson Observatory, near Los Angeles, where he studied the infrared radiation of the Sun.

Awarded his doctorate in physics in 1922 by Johns Hopkins University, Brackett continued his studies of hydrogen using gas discharge tubes.

Through applying a high voltage across a hydrogen filled gas discharge tube, Brackett discovered the series of lines resulting from an electron transitioning from a higher energy level to the fourth level. There are a series of lines in the Brackett which begin at 4.05 μm and converge into a continuum, the series limit, at 1.51 μm and can be calculated from the *Rydberg equation*:

$$\frac{1}{\lambda} = R_H \left(\frac{1}{n_1^2} - \frac{1}{n_2^2} \right)$$

where λ is the wavelength (in metres), $R_H = 1.097376 \times 10^{-7}$ m^{-1}, $n_1 = 4$ and $n_2 = 5, 6, 7 \ldots$

That the transition was to the fourth level was indicated by the fact that the series of lines could only be calculated correctly when $n_1 = 4$ and $n_2 > 4$. The three most significant are the transitions between $n_1 = 4$ and $n_2 = 5$, 6 and 7 and the convergence occurs when $n_2 = 18$. This indicated that the hydrogen has eighteen possible energy levels in which the electron can reside before being ejected when provided with even more meaning that the hydrogen atom is then ionised.

Before the Second World War Brackett taught at Berkeley before joining the Department of Agriculture's Fixed Nitrogen Lab in 1927.

In 1929, Brackett left academia to join the Fixed Nitrogen Laboratory of the United States Department of Agriculture and seven years later transferred to the National Institutes of Health (NIH) where he directed a department working on applying spectroscopic methods to detect and measure chemicals in body fluids.

During World War II, Brackett served in the Army, reaching the rank of lieutenant colonel, directing a research programme designing vision and fire control optics for combat vehicles.

In 1947, he was appointed to lead the Section of Photobiology in the National Cancer Institute's (NCI) Laboratory of Physical Biology and it was there that he introduced computing and developed techniques to interface computers with scientific instruments.

After peace was declared, Brackett returned to the National Institutes of Health and directed the photobiology section until his retirement in 1961.

In recognition of his spectroscopic achievements, the lunar crater *Brackett* is named after him.

William Henry Bragg

 Sir **William Henry Bragg** OM PRS (1862 – 1942) was a physicist, chemist, mathematician and sportsman. Bragg and his son, Sir William Lawrence Bragg were joint recipients of the 1915 Nobel Prize in Physics *'for their services in the analysis of crystal structure by means of X-rays'*. Born in Cumberland, Bragg was raised by his uncle in Leicestershire after the death of his mother when he was seven. He studied won an exhibition (scholarship) to Trinity College, Cambridge to study mathematics, graduating with first class honours.

At the astonishingly young age of twenty three, Bragg was appointed (Sir Thomas) Elder Professor of Mathematics and Experimental Physics in the University of Adelaide. The university's facilities were very basic and Bragg apprenticed himself to a firm of instrument makers in order to learn how to produce the necessary equipment.

Bragg's interest in physics developed when he was visited by Lord (Ernest) Rutherford who was moving from his native country of New Zealand to work at the Cambridge Cavendish Laboratory. Bragg became fascinated by electromagnetism and by the newly discovered x – rays and he was the first in Australia to publicly demonstrate the use of this radiation to reveal structures of the body mainly by performing x – ray analysis of his own finger and his son's (William Lawrence Bragg) broken arm. Simultaneously, Bragg was working on electrical telegraphy and gave the first public demonstration of wireless telegraphy (radio) in Australia. Bragg came to fame in 1904 when he gave a series of lectures on radioactivity and on the theory of the ionisation of gases and published what became a standard book '*Studies in Radioactivity*'. After twenty three years in Australia, Bragg returned to Britain in 1908 on the maiden voyage of the SS Waratah which vanished on its second journey. Appointed to the Cavendish chair of physics at the University of Leeds from 1909 he invented the x – ray spectrometer and with his son, Lawrence, created the discipline of x – ray crystallography. This led to the '*Bragg Controversy*' about sodium chloride. The Braggs determined that sodium chloride, until then believed to be a molecule, was actually a three – dimensional crystalline material with alternating sodium and chloride ions. This caused a huge fuss as the concept of an ionic crystalline material was anathema to those who could not believe that compounds could exist except as discrete molecules.

In 1914, Bragg was appointed Quain Professor of Physics at University College London (UCL) and in 1915 was appointed to the Admiralty's Board of Invention and Research where he worked on sonic measurements to detect German U – Boat submarines, the three dimension geolocation of the sites of explosions i.e. massive German weapons, to determine the correct location of British warships and to locate minefields. After the Armistice, Bragg returned to UCL and in 1923 was appointed Fullerian Professor of Chemistry at the Royal Institution and director of the Davy Faraday Research Laboratory.

Bragg delivered the world famous Royal Institution Christmas Lectures (instituted by Michael Faraday) in 1919, 1923, 1925 and 1931 on the diverse subjects on '*The World of Sound*' (1919); '*Concerning the Nature of Things*' (1923), '*Old Trades and New Knowledge*' (1925) and '*The Universe of Life*' (1931).

Elected President of The Royal Society in 1935, subsequently Bragg became an advisor to the British government, a position he retained until his death in 1942.

William Lawrence Bragg

Sir **William Lawrence Bragg**, CH, MC, FRS (1890 – 1971) was the Australian – born son of Sir William Henry Bragg and was joint recipient, with his father, of the 1915 Nobel Prize in Physics for *'Their services in the analysis of crystal structure by means of X-rays'*.

Soon after starting school, Bragg fell off his tricycle and broke his arm. His father examined the broken arm using x – rays to determine the exact position of the break in the bone. This was the first use of x – ray radiography in Australia. Following his father's appointment to the Cavendish Chair of Physics at the University of Leeds, Bragg studied mathematics at Trinity College, Cambridge but changed to physics for his final years before becoming a physics research student.

It was at the Cavendish Laboratory that he began to investigate x – rays. There was huge controversy at the time since many eminent scientists claimed that the radiation was simply electromagnetic radiation whilst other, equally eminent, physicists had decided that they must be particulate in nature. The matter was resolved by Max von Laue who, in an experiment analogous to Thomas Young's visible diffraction experiment, determined that x-rays directed at a crystal produced spots on a photographic plate which could not have been produced by particles.

Bragg's insight was to determine that crystals comprised of parallel plates of atoms or ions would not diffract the x – rays since the diffracted waves would destructively interfere and cancel each other out. On the other hand, if the waves were deflected at an angle such that the waves interfered constructively, then if the distance that they travelled between atoms equalled the wavelength of the x – rays then spots would be detected on the photographic plate. From this stunning insight, he wrote the very simple Bragg equation:

$$n\lambda = 2\,d\,\sin\theta$$

where n is an integer, λ is the wavelength of the x – ray radiation, d is the distance between the atoms or ions and θ is the angle of diffraction. Using his father's x – ray spectrometer, Bragg was able to calculate the distance between atoms or ions in a crystalline material.

During the First World War, Bragg and his brother, Robert, were commissioned in to the army. Lawrence was seconded to the Royal Engineers to work on the use of sonic location of explosions. Robert was killed in the Gallipoli campaign. Whilst in service, Bragg and his father were jointly named as Nobel Laureates in Physics for 1915. Bragg was twenty five years old and remains, to this day, the youngest ever Nobel Physics laureate. In 1938, he became Professor of Natural Philosophy at the Royal Institution and delivered the Faraday Christmas lectures in 1934 and 1961.

Between the wars, Bragg was Langworthy Professor of Physics at Manchester (replacing Ernest Rutherford) before becoming the director of the National Physical Laboratory in 1937. In 1938, he replaced Rutherford as the Cavendish Professor of Physics and also became director of the Laboratory of Molecular Biology, in which role he was serving when Watson and Crick reported their discovery of the structure of DNA.

Louis de Broglie

Louis Victor Pierre Raymond de Broglie (1892 – 1987), 7th duc de Broglie, postulated in his, 1924, doctoral thesis, that electrons like all matter have the properties of both waves and particles and the concept became known as *wave – particle duality*. It formed the foundation of the theory of quantum mechanics and de Broglie won the 1929 Nobel Prize in Physics after the wave properties of electrons was demonstrated (1927) had been been experimentally verified.

He proposed that, in addition to Planck's relation of energy (E) and frequency (ν)

$$E = h\nu$$

where h is Planck's constant, 6.636 x 10^{-34} Js^{-1}

the wavelength of an electron (λ) is related to its momentum (p) by the relation:

$$\lambda = h / p$$

de Broglie was a theoretician and not an experimentalist and his theory had to await validation by others. This was achieved almost simultaneously in Britain, Sir G.P. Thomson, and in America, by Davisson and Germer.

Working at the University of Aberdeen, George Paget Thomson passed a beam of electrons through a thin film of celluloid and observed the predicted interference patterns whilst at around the same same Clinton Joseph Davisson and Lester Halbert Germer, at the Bell Labs, diffracted a beam of electrons from the surface of nickel and produced a diffraction pattern predicted by Thomson for x-rays. Thomson and Davisson shared the 1937 Nobel Prize in Physics for their, independently made, discoveries. de Broglie's wave – particle duality model was the basis for the formulation of quantum mechanics by Erwin Schrödinger.

de Broglie was a scion of the aristocratic de Broglie family who for centuries had occupied military, scientific and political positions. One of his brothers, Maurice, 6th duc de Broglie became a famous experimental physicist studied X-ray diffraction, spectroscopy and radio. Their father had been the 5th duc de Broglie and, on his brother's death, Louis succeeded Maurice in that elevated position because Maurice's only son had died in infancy. Both Maurice and Louis were members of the prestigious Académie française and the Académie des sciences.

Louis de Broglie had initially tended towards the humanities and took his first degree in history before turning to physics whilst in the First World War, he served in the French army working on military radio communications.

In addition to his strictly theoretical work, de Broglie wrote extensively about the philosophy of science.

He also established the Centre for Applied Mechanics at the Henri Poincaré Institute, whose researchers concentrated on optics, cybernetics and nuclear energy.

Robert Bunsen

Most famous for his laboratory burner, Göttingen – born **Robert Wilhelm Eberhard Bunsen** (1811 – 1899), investigated the emission spectra of heated elements, and working with Gustav Kirchhoff, discovered caesium (1860) and rubidium (1861).

Interested in geochemistry, physics and physical chemistry, Bunsen became a lecturer at Göttingen and, working on the chemistry of arsenic, developed a method of precipitating arsenic which is still used today for the treatment of arsenic poisoning.

His studies of arsenic brought him fame for his experimental skills since arsenic and its compounds are extremely toxic and very difficult to work with. Indeed, he lost an eye when a sample of tetramethyldisarsine $(CH_3)_2As–As(CH_3)_2$, which spontaneously combusts in air, exploded.

In 1852, Bunsen became Professor at the University of Heidelberg where he developed electrolytic techniques for the production of samples of alkali and alkaline earth metals before moving to study photochemistry and he established the photochemical formation of hydrogen chloride from hydrogen and chlorine.

Starting in 1859, Bunsen worked with Kirchhoff to study the emission spectra of elements. Essentially, using his high temperature Bunsen Burner, he put as many materials as he could find in the flame and collected their spectra.

Although the colours of different metals been noted before and were essential in identifying the metals present, Bunsen and Kirchhoff were the first to collect spectra systematically using their prototype spectroscope and measured the spectra of lithium, sodium and potassium.

Other lines that could not be assigned led to the almost immediate discovery of the previously unknown elements, caesium and rubidium.

Caesium was discovered when Bunsen and Kirchhoff identified blue lines in mineral water collected near Dürkheim and after distillation of almost forty tons of the water succeeded in collecting seventeen grammes of the caesium – containing compound. Rubidium was discovered in a similar manner.

Bunsen was one of the most popular and famous scientists and teachers of his time. He was devoted to education and mentoring his students.

He also believed that scientific knowledge should not be owned by any individuals and this meant that, although he could have become a very wealthy man, he did not do so as he resolutely refused to patent any of his discoveries or inventions since he believed that all knowledge belonged to mankind and not to any particular person.

James Chadwick

Sir **James Chadwick**, CH, FRS (1891 – 1974) was awarded the 1935 Nobel Prize in Physics for his discovery of the neutron in 1932 and was heavily involved the development of nuclear weapons.

Born into relative poverty, Chadwick studied at Manchester under Lord Rutherford and, living at home, he had to walk four miles each way to attend the university for which he had won two scholarships. He had meant to study mathematics but found he had enrolled on the physics course by mistake. Graduating with First Class Honours in 1911 he spent two years as a postgraduate student and received his masters' degree in 1913 for his work on the measurement of the wavelengths and the intensity of gamma radiation. Awarded a scholarship to study the beta radiation emitted by radioactive metals in Berlin under Hans Geiger, Chadwick determined that beta radiation emitted a continuous spectrum rather than the expected discrete lines. Present in Germany at the outbreak of World War I, he spent the war years interned in Ruhleben where he was allowed to set up a laboratory in disused stables and worked on matters such as the ionisation of phosphorus and the photochemical reaction between carbon monoxide and chlorine.

After the Armistice and his release, Chadwick became a doctoral student under Rutherford again but this time at the Cambridge Cavendish Laboratory. Following completion of his post-graduate studies, Chadwick worked for a decade as Rutherford's assistant director of research and it was here that he discovered the existence and properties (neutral charge and mass) of the neutron.

Chadwick left Cambridge in 1935 when appointed to a chair of physics at the University of Liverpool and it was here that constructed a cyclotron, making Liverpool a world class leader in the study of nuclear physics. The neutron facilitated the artificial formation of elements heavier than uranium. Attempts had been made using protons but the repulsion of protons by the targeted nucleus prevented success. Neutrons, possessing spin but no charge could be absorbed by the nucleus since there was no Coulombic repulsion. The new nuclei would then decay by the emission of beta radiation implying that the neutron is not a fundamental particle but is actually comprised of a proton – electron pair. This also explained why absorption of a neutron could lead to an increase in the atomic mass of the element by the following process:

$$\text{neutron} \longrightarrow \text{proton} + \text{electron} + \text{emitted radiation}$$

$$^{0}_{1}\text{n} \longrightarrow ^{+1}_{1}\text{p} + ^{-1}_{0}\text{e} + h\nu$$

This equation explain the demonstration that the neutron made it possible to produce elements heavier than uranium in the laboratory by the capture of slow neutrons followed by beta decay. As stated earlier, unlike the positively charged alpha particles, which are repelled by the forces present in the nuclei of other atoms, neutrons do not need to overcome any Coulomb barrier, and can therefore penetrate the nuclei of even the heaviest elements such as uranium. This became a fundamental process in nuclear fission.

Chadwick's discoveries inspired Enrico Fermi to investigate the nuclear reactions brought about by collisions of nuclei with slow neutrons, work for which Fermi would receive the 1938 Nobel Prize in Physics.

There was still a problem. Chadwick's continuous spectrum of beta radiation energies could not be explained without violating the the well established *Principle of the Conservation of Energy*.

In 1930, Wolfgang Pauli proposed the existence of another kind of particle of zero charge and zero mass. If this particle was emitted then the continuous emission spectrum of beta radiation was accounted for. Pauli termed this a neutron as well but Fermi renamed it the *neutrino* for *little neutron* when he proposed a theory of nuclear decay comprising the decay of a neutron into a proton, electron, neutrino and emission of radiation.

$$\text{neutron} \longrightarrow \text{proton} + \text{electron} + \text{neutrino} + \text{emitted radiation}$$

$$^{0}_{1}n \longrightarrow {}^{+1}_{1}p + {}^{-1}_{0}e + {}^{0}_{0}\nu + h\nu$$

The theory accounted for the missing mass but even now, in 2020, the neutrino is incredibly difficult to detect since, having no mass and no charge, they are theoretically capable of passing through the Earth completely undetected.

During the Second World War, Chadwick was heavily involved in the project to build a nuclear weapon whilst all the time the facilities were being bombed by the Luftwaffe.

When Britain and America combined their research capabilities, Chadwick moved to Los Alamos and then Washington D.C. and was an observer of the first explosion of a nuclear bomb. He was also present when the British government signed an agreement with America to use the atomic bomb against Japan.

Chadwick found it incredibly difficult to reconcile his discovery of the neutron with the lethality of atomic weapons and started using sleeping tablets which he continued using for the rest of his life.

Chadwick and his family returned to Britain in 1946.

The then Vice Chancellor of the University of Liverpool, Sir James Mountford, recorded in his diary that he [Mountford] had never seen a man [Chadwick] so *physically, mentally and spiritually tired for he had plumbed such depths of moral decision as more fortunate men are never called upon even to peer into and suffered almost insupportable agonies of responsibility arising from his scientific work*.

In 1948, Chadwick became Master of Gonville and Caius College before finally retiring in 1958.

Clinton Davisson

Illinois – born **Clinton Joseph Davisson** (1881 – 1958) was awarded the 1937 Nobel Prize in Physics for his discovery of electron diffraction through the famous Davisson-Germer experiment. Davisson shared the Prize with George Paget Thomson, who independently discovered electron diffraction at the same time.

Educated at the University of Chicago he was subsequently appointed in 1905, to the physics department at Princeton on the recommendation of Robert Millikan who was awarded the 1923 Nobel prize in Physics for his elegant experiment which determined the charge on the electron.

Subsequently appointed to the Carnegie Institute of Technology and working during the war years in the Engineering Department of the Western Electric Company (later the Bell Telephone Laboratories), Davisson was appointed there permanently and he remained with the company for the remainder of his career since he was more interested in research than teaching.

In 1927, Davisson and Lester Germer performed the classic electron diffraction experiment at the Bell Labs which verified Louis de Broglie's *wave – particle duality* theory which involved aiming slow moving electrons at a crystalline nickel target.

Ironically the aim of the experiment was not to study electrons but to analyse crystalline, metallic nickel. The diffraction pattern of the diffracted electrons perfectly matched the pattern that the Braggs had predicted for x-rays. That waves, *x – rays* could behave as particles and electrons, supposedly *particles* could behave as waves accidentally verified the wave – particle duality of matter.

Moreover, the measured wavelength agreed well with de Broglie's equation

$$\lambda = h \,/\, p$$

where the wavelength of an electron (λ) is related to its momentum (p) and h is the Planck constant, which equals 6.626×10^{-34} J s^{-1}.

Davisson found teaching interfered too much with his research and so remained at Western Electric (and Bell Telephone) until his formal retirement in 1946.

He then accepted a research professor appointment at the University of Virginia that continued until his second, and final, retirement in 1954

Albert Einstein

Everybody who has ever heard of **Albert Einstein** (1879 – 1955) has an image of him as a mad scientist with wild white hair. The hair may have developed in later life but when he rose to fame for his explanation of the photoelectric effect and for much of his life, he was very smartly dressed and well presented as shown right.

Everybody knows of the famous equation

$$E = mc^2$$

where E = energy, m is the mass of a particle and c is the speed of a light.

This is not exactly a correct statement since c is the velocity of all mass-less particles in a vacuum of which the photon of light is only one example.

They also all know about his foundation work on the theory of special and general relativity but it was for explanation of the *photoelectric effect* that he was awarded the 1921 Nobel Prize in Physics.

This has already been discussed earlier in the volume and was of enormous significance in that he related the theoretical Planck theory of quanta – packets of energy – to emission spectra and led directly to Niels Bohr's theory of the structure of the hydrogen atom.

German – born, but who took Swiss nationality in 1896, Einstein's most productive year was 1905 when he published four papers.

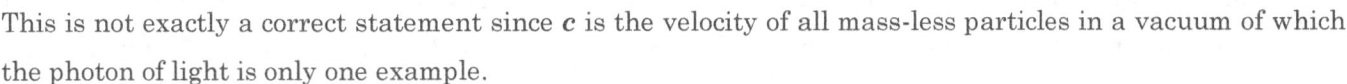

- The first was explained the photoelectric effect,
- The second paper explained Brownian motion and proved the existence of the atom whose existence had been disputed by some for many years.
- His third paper introduced special relativity, and
- The fourth described mass-energy equivalence.

Einstein was in America when Hitler came to power and, being Jewish, he never returned, taking American citizenship in 1940.

In 1939 he wrote to President F.D. Roosevelt warning that Germany would begin to develop nuclear weapons. This led directly to the American *Manhattan Project* but for the rest of his life, Einstein was opposed to nuclear weapons.

In the 1950s Einstein became perhaps the first scientific media star of the modern era and it was only then he played up to his, fictional, mad white – haired scientist, image.

Edward Frankland

Catterall – born, Sir Edward Frankland (1825 — 1899) was one of the most significant British chemists of the 19th century and is renowned as *'the father of organometallic chemistry'*, developed the concept of valency through which he created the foundations of structural chemistry and also became an expert in water chemistry.

He was born illegitimate after his mother, Peggy Frankland, a servant of the wealthy Gorst family of Preston, had an affair with the son and heir Edward. Horrified by the potential scandal and public disgrace, the Gorst family dismissed her but paid her an annuity to ensure her silence. Peggy latter married a cabinet maker, William Helm, and used the annuity to fund Edward's schooling in Manchester, Salford and then Lancaster where he became interested in chemistry after reading the writings of Joseph Priestley and hearing about the prosecution of a chemical manufacturer for a leak of hydrochloric acid which was then known as muriatic acid.

Initially wishing to become a doctor but prevented by the cost, he was apprenticed to an apothecary at the age of fourteen. During his five-year apprenticeship, Frankland began to learn to conduct chemical experiments and also attended laboratory classes at the Lancaster Mechanics' Institute run by a local doctor, James Johnson, who secured him a place in the Westminster laboratory of Dr. Lyon Playfair, 1st Baron Playfair, who had studied under Justus Liebig. Playfair secured him a place at the University of Marburg, where he studied under Robert Bunsen and from where he graduated with a doctorate before moving to Giessen to train with Liebig. It was at Marburg that Frankland performed the first laboratory synthesis of an organometallic chemical when, he synthesised both diethylzinc and dimethylzinc by reacting, individually, ethyl iodide and methyl iodide with metallic zinc.

In 1851, Frankland became the first professor of chemistry at Owens College, a constituent college of the University of Manchester and then succeeded Playfair as professor of chemistry at the College for Civil Engineers where he began the formulation of the principles of organic chemistry and conceived the cept of valency which underpins all structural organic chemistry. In 1863, Frankland succeeded Michael Faraday as professor of chemistry at the Royal Institution and two years later also became a professor of chemistry at the Royal School of Mines.

In his, 1866, Lecture Notes for Chemical Students, Frankland formalised the concept of valency, was the first chemist to represent chemical elements by letters and introduced the concept of chemical bonds which he represented with the, now familiar, straight line – for bonds. Continuing his work on valency, Frankland also introduced the concept of carbon – carbon double and triple bonds, represented by $=$ and $\equiv$, respectively.

In 1851, Frankland married Sophie Fick, who died of tuberculosis in 1874 and, in 1875, he married Ellen Grenside. Frankland died, in 1899, whilst on holiday in Norway, leaving his widow and a son, Percy, who was also an eminent chemist with expertise in water pollution and fermentation and who was the godson of Michael Faraday.

Joseph Fraunhofer

Bavarian – born **Joseph Ritter von Fraunhofer** (1787 – 1826) was a physicist and optical lens manufacturer.

Orphaned aged eleven, he worked for a glassmaker and was almost killed when the workshop collapsed. He was rescued in an operation led by Prince-Elector Maximilian Joseph who subsequently supported Fraunhofer financially and provided him with books and time to study.

Having become expert in the manufacture of optical glass and telescope lenses, Fraunhofer invented the spectroscope, and developed diffraction gratings. He is pictured below demonstrating his spectroscope.

Fraunhofer worked with a number of partners in various optical businesses and he became Director of the prestigious Optical Institute

Fascinated by sunlight he was the the first to study it practically and discovered the dark absorption lines in the spectrum of sunlight which are now known as Fraunhofer lines.

In 1859, Kirchhoff and Bunsen demonstrated the dark fixed lines that were later shown to be atomic absorption lines.

Glassmaking involved the use of mercury and, as with many other glassmakers, Fraunhofer suffered from mercury poisoning before dying of tuberculosis at the tragically young age of thirty nine.

Joseph Gay-Lussac

The Haute-Vienne – born chemist and physicist, **Joseph Louis Gay-Lussac** (1778 – 1850), is most famous for discovery that the formula of water is H_2O. Until then it was believed to be HO.

Educated at the École Polytechnique and then the École des Ponts et Chaussées, Gay–Lussac was appointed professor of chemistry at the École Polytechnique before becoming a professor of physics at the Sorbonne.

In 1809, Gay-Lussac married a shop assistant, Geneviève-Marie-Joseph Rojot, who he discovered was spending her free time studying a chemistry textbook under the counter. They had five children, the eldest of whom, Jules, became an assistant and student of Justus Liebig in Giessen.

In 1802, he formulated his gas law that related pressure to temperature and resulted from his travels in a hot air balloon collecting samples of air to measure their pressure and temperature. To conduct this experiment, Gay – Lussac and Jean-Baptiste Biot ascended to a height of over seven kilometres, collecting air samples at different heights.

Following his 1802 experiments, Gay-Lussac's stated the gas law

$$P = k\,T$$

where P = pressure, T = temperature and k is a constant of proportionality

This laid the basis for the ***Universal Gas Law***,

$$pV = nRT$$

where p = pressure, T = temperature, V = volume, n = number of moles and R is the Universal Gas Constant, 8.314 $JK^{-1}\,mol^{-1}$, where the temperature is recorded in Kelvin.

In a similar experiment in 1805, and working with Alexander von Humboldt, he determined that, although atmospheric pressure does change with altitude, its composition does not vary with altitude.

In 1808, he was one of several chemists who discovered the existence of the element, boron and three years later, discovered and named the element, iodine.

In 1824 he also reported the development of the *pipette* (used to dispense fixed volumes) and *burette* (used to measure dispensed volumes and coined the name for both essential items of laboratory glassware.

Hans Geiger

Neustadt an der Haardt – born **Johannes Wilhelm *Hans* Geiger** (1882 – 1945) is best known as the co-inventor of the Geiger – Muller radioactivity counter and for the Geiger–Marsden experiment which, conducted under Lord (Ernest) Rutherford, discovered the existence of the atomic nucleus.

Following the award of his doctorate from the University of Erlangen, in 1906, on electrical discharges through gases he was appointed to a fellowship at the University of Manchester where, under Rutherford, he worked with Ernest Marsden and conducted the famous '*gold foil experiment*' which determined the existence of the nucleus and led to Rutherford winning the 1908 Nobel Prize in Chemistry.

In 1912, Geiger became head of radiation research at Berlin's German National Institute of Science and Technology where he worked with James Chadwick.

During the First World War Geiger served in the German army as an artillery officer.

In 1924, Geiger confirmed the existence of the Compton effect (scattering of photons by electrons led to an increase in the wavelength of the impinging radiation) which confirmed the wave – particle duality of light and in 1929 became Professor of Physics at the University of Tübingen before moving to the Technische Universität Berlin in 1936.

He is most well known for the Geiger – Muller tube for detecting radiation which can be adapted to detect all of alpha, beta, gamma, x-radiations and neutrons and is well known for the sounds produced when such radiation is detected.

In the early years of the Second World War he was involved in the feasibility study of nuclear weapons (conducted by the '*Uranium Club*') which he opposed and he died two months after the first nuclear bomb exploded over Hiroshima.

Lester Germer

Chicago – born **Lester Halbert Germer** (1896 – 1971) performed the famous Davisson – Germer electron diffraction experiment which demonstrated the wave-particle duality of matter and which led to the development of the electron microscope.

In the First World War, Germer was a fighter pilot and after the Armistice moved to the Bell Laboratories in New Jersey where he worked with Davisson and also investigated thermionic emission (the emission of electrons by heated metals) and the corrosion of metals.

The Davisson – Germer experiment was not intended to investigate electrons rather the electrons were intended to be used to investigate the structure of crystalline nickel. In a vacuum, electrons from a heated filament were targeted at the nickel target which could be rotated as could the detector, a Faraday Cage which is a mesh covered region which prevents the transmission of electromagnetic waves.

The experimental arrangement is shown to the right.

To their surprise they measured sudden intensities in the diffracted electrons which indicated, using the Bragg equation (below) the distance between the nickel ions in the crystalline lattice.

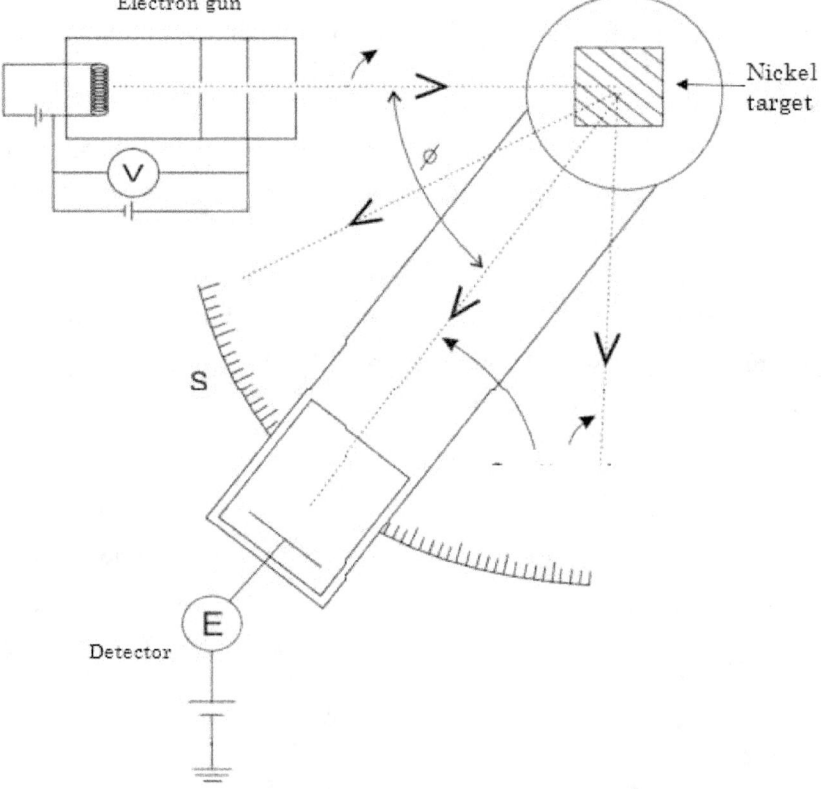

$$n\lambda = 2\, d\, \sin\theta$$

This diffraction demonstrated that the electrons were behaving like waves and the measurements, in combination with the Bragg equation (above) could determine the distance (d) between the ions and is one of the most important experiments underlying the entire theory of quantum mechanics.

A keen rock climber, Germer suffered a fatal heart attack when climbing on the Shawangunk Ridge in New York State.

Eugen Goldstein

Eugen Goldstein (1850 – 1930) born in Gliwice, which is now in Poland, was one of the first to investigate the behaviour of gases in gas discharge tubes.

Goldstein studied at Breslau before moving to Berlin to study under Hermon von Helmholtz, in Berlin. Goldstein worked at the Berlin Observatory from 1878 to 1890 but became head of astrophysics at the Potsdam Observatory in 1927.

In the mid-nineteenth century, Julius Plücker had investigated the light emitted in discharge tubes (the Crookes tubes) and the effect of magnetic fields on the emitted light. In 1869, Johann Wilhelm Hittorf studied discharge tubes with energy rays extending from a negative electrode, the cathode. He observed that these rays produced a fluorescence when they hit the tube's glass walls. When interrupted by a solid object they cast a shadow.

In the 1870s, Goldstein studied the radiation from discharge tubes and named the emissions cathode rays, *Kathodenstrahlen* contributing to J.J. Thomson's discovery of the electron. Specifically, he observed that cathode rays were emitted perpendicularly from a metal surface, and carried energy.

In 1886, he discovered that tubes with a perforated cathode also emit a glow at the cathode end.

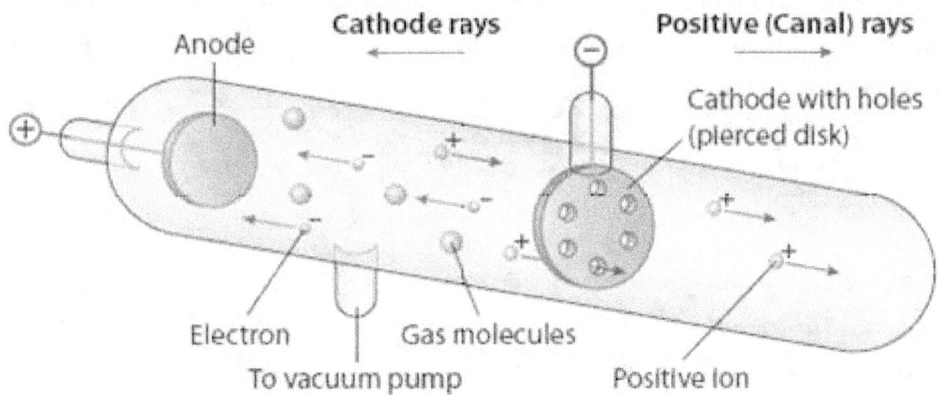

Goldstein concluded that, in addition to cathode rays, which travelled from the (negatively charged) cathode toward the (positively charged) anode, another ray travelled in the opposite direction. Since these rays passed through the holes in the pierced cathode disk, known as channels or canals, Goldstein called them canal rays, *Kanalstrahlen*.

It was clear that these rays were attracted to the negative electrode and had to carry a positive charge. Goldstein had discovered the *proton*.

An astrophysicist by profession, Goldstein also used the discharge tubes to investigate comets and determined that a small object such a ball bearing or glass ball scattered the radiation and gave the appearance of a comet's tail.

Cornelis Gorter

The Utrecht – born **Cornelis Jacobus *'Cor'* Gorter** (1907 – 1980) discovered *paramagnetic relaxation*, the form of magnetism where some materials are weakly attracted by an externally applied magnetic field. Such materials form induced magnetic fields which act in the direction of the applied magnetic field. The relaxation occurs when the magnetic field is removed or simply, in the case of an electromagnet is simply switched off. Gorter describe this as a resonance effect and may thus be regarded as one of the fathers of nuclear magnetic resonance spectroscopy. He was also a pioneer in the study of superconductivity and reconciled the phenomenon with thermodynamics and with the equations of James Clerk Maxwell.

Gorter was a true general physicist, rejected specialisation and regarded the whole subject to be accessible. Appointed to Leiden after the Second World War he devoted much time and energy to develop its laboratories into the some of the most modern and sophisticated facilities in the world.

Paramagnetic relaxation is a fundamental principle underlying the discipline of nuclear magnetic resonance but he did not invent it; he discovered it. Through his cryogenic studies he theorised about low temperature electrical resistance of metallic films and came close to discovering the technique of electron tunnelling.

A very modest man, he was awarded the *Fritz London Award* in 1966 and his acceptance speech was titled *'Bad luck in attempts to make scientific discoveries'*, in which he discussed how many times he had just missed out on making a number of such important discoveries.

Gorter published a monograph *'Paramagnetic Relaxation'* in 1946 and wrote a moving introduction relating the appalling conditions that the Dutch lived under during the German occupation:

'The greater part of the present monograph was written during the winter of 1944-45 known in Holland as "starvation winter". At that time the densely populated Western part of the Netherlands was cut off from the South by the fighting-line and from the East by a broad zone of German military posts. Owing to the German drives for slave-workers it was often risky for men under forty to go about in the streets; only very urgent duties and the necessity to procure food or wood for fuel could induce them to leave their houses. In spite of this, however, scientific work was continued here and there. In the Zeeman Laboratory of the University of Amsterdam enough fuel was left for one room to still be heated, so that this building remained one of the few centres where research work was carried on. The news supplied clandestinely by the radio, run on the batteries of the laboratory, constituted also an attraction to the scientific and technical personnel. Owing to the absence of electricity and gas activities were mostly of a theoretical nature: writing theses, discussing theoretical problems, designing, calculating and working out previous observations. Though from an objective point of view the value of this work was perhaps not outstanding, it helped people to rid themselves for a time of the daily obsession and anxiety about food, warmth and the slow progress of the war and to hold their own as self-respecting scientific workers. After the Liberation, there was so much other work to do that it took some time before the last chapter was written and the whole had been revised.'

C. J. Gorter, Leyden, November 1946

Nagaoka Hantaro

Nagasaki – born **Nagaoka Hantarō** (1865 – 1950) graduated with a physics degree from the University of Tokyo in 1887 and almost immediately started working with a visiting Scottish physicist, Cargill Gilston Knott, on magnetism. In 1893, Nagaoka moved to Europe, where he studied at the Universities of Berlin, Munich, and Vienna.

Significantly for his future work, Nagaoka attended lectures on the rings of the planet, Saturn and worked with Ludwig Boltzmann on the latter's *Kinetic Theory of Gases*. In 1900, Nagaoka also attended the *First International Congress of Physicists* in Paris. It was here that he heard Marie Curie lecture on radioactivity and her lecture aroused his interest in atomic physics.

In 1901, Nagaoka returned to Japan and was appointed professor of physics at the University of Tokyo. He subsequently served as President of Osaka University from 1931 to 1934.

For the purposes of this volume, Nagoaka's major contribution was to propose his *Saturnian model of the atom*.

Since the late 19th century, physicists had learned that the atom could no longer be considered to be a hard, solid and impenetrable particle. This had arisen from Goldstein's discovery of the *proton* and Thomson's discovery that cathode rays were the negatively charged particles that we now term *electrons*.

In 1903, Thomson proposed his '*plum pudding*' model of the atom which suggested that the atom was a sphere of overall neutral charge with positive charges and, negatively charged, electrons scattered throughout it. Nagaoka rejected Thomson's model and proposed an alternative planetary model of the atom in which a positively charged centre is surrounded by a stream of revolving electrons rather like the planet Saturn and its rings. Nagaoka's model made two predictions:

- The atom comprises a dense but massive atomic centre, with;
- Electrons revolving around the nucleus, bound by electrostatic forces, analogous to the rings revolving around Saturn, those bound by gravitational forces.

Both predictions were confirmed by Lord (Ernest) Rutherford, in his (1911) paper on the Geiger – Marsden experiment which he mentored and planned. In fact, Rutherford even referred to Nagaoka's model when he demonstrated the existence of the atomic nucleus. Nagoaka however abandoned his model when it was calculated that the stream, or train, of electrons would repel each other and could not exist as a stream in one simple set of orbiting electrons.

Werner Heisenberg

 Bavarian – born Werner Heisenberg (1901 – 1976) rose to fame in the world of physics in 1925 with the first of a series of seminal papers on quantum mechanics. He established a matrix formulation of quantum mechanics and is renowned for the *Heisenberg Uncertainty Principle* and he was awarded the 1932 Nobel Prize in Physics '*for the creation of quantum mechanics*'.

Heisenberg also contributed to the theories of the hydrodynamics of turbulent flows, ferromagnetism, the atomic nucleus and cosmic rays and was one of the principal scientists in the German nuclear weapons programme and met Niels Bohr in German - occupied Copenhagen.

That he was involved in the German weapons' programme is fascinating since, after Hitler's rise to power, he was attacked in the German press as a '*white Jew*' for teaching Einstein's theory of relativity in his university course and criticised by anti-Semitic scientists who were also critical of the entire concept of theoretical physics. This was resolved by Himmler whose mother knew Heisenberg's mother through their husbands who were both ministers and members of the same hiking club. Himmler sent, simultaneously, letters to Reinhard Heydrich and Heisenberg. In his letter to Heydrich, Himmler wrote that Germany could not afford to lose or silence Heisenberg and to Heisenberg he advised him to distinguish between the theories and political opinions of the scientists who had criticised him.

Heisenberg's matrix mathematics is beyond the scope of most people, including most mathematicians, but it is the conclusions that are important. Essentially, he concluded that reality is a philosophical concept and although we can speak of mathematical knowledge we cannot have any 'true' access to the particles themselves. With regard to the electron, the consequences are that we cannot, simultaneously, know both its momentum and location.

A further consequence is that we can no longer refer to the behaviour of the particle independently of the observation and that the mathematically formulated natural laws of quantum theory do not deal with the particles themselves but simply describe our knowledge of them. This led to the conclusion that we cannot picture nature as it really exists but we can only talk of our perception of it. He also wrote that '*Nor is it any longer possible to ask whether or not these particles exist in space and time objectively ... When we speak of the picture of nature in the exact science of our age, we do not mean a picture of nature so much as a picture of our relationships with nature ... Science no longer confronts nature as an objective observer, but sees itself as an actor in this interplay between man [sic] and nature. The scientific method of analysing, explaining and classifying has become conscious of its limitations, which arise out of the fact that by its intervention science alters and refashions the object of investigation. In other words, method and object can no longer be separated.*' This is a modern example of why physics, as we are taught it, was previously known as Natural Philosophy.

Heisenberg was a talented pianist and, aged thirty five he met his future wife, Elisabeth Schumacher, at a concert at which he performed. They became engaged two weeks later and remained together until his death. The devoted couple, pictured above, had seven children including three who became very eminent: the geneticist, Martin Heisenberg; educational psychologist, Christine Mann and the nuclear physicist, Jochen Heisenberg.

Heinrich Hertz

Hamburg – born **Heinrich Rudolf Hertz** (1857 – 1894) conclusively proved the existence of the electromagnetic waves predicted by James Clerk Maxwell and established the foundations for Marconi's development of radio transmissions and reception.

Hertz was equally adept at both science and languages and learned to speak and write both Arabic and Sanskrit. He studied physics and engineering in Dresden, Munich and Berlin. In Berlin he studied Gustav R. Kirchhoff and Hermann von Helmholtz.

After receiving his doctorate in 1880 from the University of Berlin he spent three years as a post-doctoral research student under Helmholtz before becoming a lecturer in theoretical physics at the University of Kiel and, in 1885, a professor of physics at the University of Karlsruhe.

It was at Karlsruhe that he conducted his experiments demonstrating the existence and properties of electromagnetic waves and his apparatus produced and detected very high frequency (VHF) radio waves. Hertz also demonstrated the polarisation of radio waves: the antenna must be in the plane of the waves which is why TV aerials must be adjusted to match the plane of the radiated waves. However, Hertz did not appreciate the importance of his discovery or the practical application, commenting that '*this is just an experiment that proves Maestro Maxwell was right — we just have these mysterious electromagnetic waves that we cannot see with the naked eye. But they are there.*' and when asked about the practical application of his discoveries his response was '*Nothing, I guess.*'.

Nevertheless, Hertz's proof of the existence of airborne electromagnetic waves led to a huge explosion of experimentation with what were termed '*Hertzian waves*' until around 1910 when they became termed '*radio waves*' and led to the 1909 Nobel Prize in Physics being awarded, jointly, to Ferdinand Braun and Guglielmo Marconi.

Hertz also established the photoelectric effect subsequently explained by Albert Einstein in 1905 (1921 Nobel Prize in Physics) when he observed that a charged object loses its charge when irradiated with ultraviolet radiation and reported these discoveries in 1887.

In 1892, Hertz suffered from a series of severe migraines and died of an infection, after numerous operations, at the tragically young age of thirty six. Although of Jewish extraction, Hertz considered himself a Lutheran Christian. That did not prevent the Nazis from removing all references to his name in the institutes in which he had worked and his widow and daughters left Hamburg for England with Hitler's rise to power.

Neither of Hertz' daughters married so he has no direct descendants but one daughter, Mathilde, became a world renowned biologist and psychologist whilst his nephew, Gustav Ludwig Hertz, received the 1925 Nobel Prize in Physics for his work on the inelastic collisions of electrons in gases. Gustav's son Hermann Gerhard Hertz, who became a professor of physics at the University of Karlsruhe, was a pioneer of nuclear magnetic resonance (NMR) spectroscopy and devised the medical technique of ultrasound investigations and treatments.

Friedrich Hund

Karslruhe - born Friedrich Hermann Hund (1896 – 1997) is renowned for *Hund's rules* which describe the multiplicity of the electronic configuration within atomic and molecular orbitals.

Essentially, Hund's rules build on the Aufbau Principle which states that electrons occupy the lowest possible energy level. Hund expanded this to state that, when there are a set of mutually degenerate orbitals such as the p - orbitals they will be occupied by a single electron each before being forced to pair up. This means that the electronic configurations of boron, carbon and nitrogen allow the three 2p – orbitals to be occupied by single electrons before they are paired up in oxygen, fluorine and neon as discussed in Chapter V.

Hund studied mathematics, physics, and geography at Marburg and Göttingen. Following his lectureship in theoretical physics at Göttingen (1925) he became a Professor at the University of Rostock (1927) and then at Leipzig University in 1929 before moving to Jena in 1947, Frankfurt/Main in 1951 and he returned to Göttingen in 1957. Hund also worked closely with Niels Bohr and lectured at Harvard in 1928.

He was a prolific writer of more than 250 research papers and worked with such eminent physicists as Erwin Schrödinger, Paul Dirac, Werner Heisenberg, Max Born and Walter Bothe and was so highly regarded that Robert S. Mulliken, awarded the 1966 Nobel Prize in Chemistry for his work on molecular orbital theory, acclaimed Hund's work and lamented that they did not share the Nobel prize which he would have welcomed to share with him.

Hund lived to the grand old age of 101 and survived long enough to appreciate how great his contribution had been to both physics and chemistry. The photograph below shows Hund (left) conversing with Werner Heisenberg (middle) and Max Born (right) at Göttingen in 1966.

Gustav Kirchhoff

Königsberg – born **Gustav Robert Kirchhoff** (1824 – 1887) made enormous contributions to the fundamental understanding of electrical circuits, spectroscopy, thermochemistry and to the emission of black-body radiation by heated objects and it was he who termed the phrase '*black-body radiation*' in 1862.

His interests were so diverse that there are several different sets of concepts, termed '*Kirchhoff's laws*' named after him.

Graduating from the Albertus University of Königsberg in 1847, Kirchhoff moved to Berlin as a researcher and lecturer until appointed to a professorship at Breslau.

Kirchhoff formulated his electrical circuit laws calculating the currents in series and parallel circuits whilst still a student and these laws became the subject of his doctoral thesis.

In 1854, he went to the University of Heidelberg where he collaborated in spectroscopy work with Robert Bunsen and invented the spectroscope with which they discovered the previously unknown elements, caesium and rubidium.

The spectroscope diagram and description shown below is taken from their paper in the prestigious journal *Annalen der Physik und der Chemie*, Vol. 110 (1860), pp. 161-189.

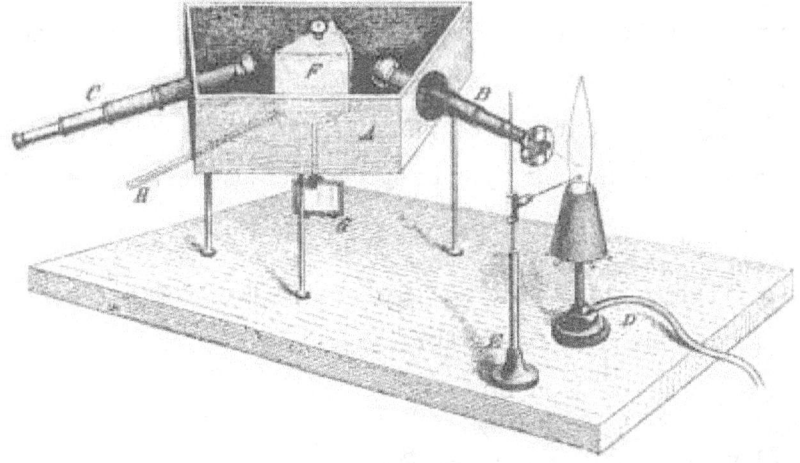

where A is an internally blackened box with a trapezoidal bottom resting on three legs; the two oblique side walls, which form an angle of about 58° with each other, carry the two small telescopes B and C. The lamp D is arranged before the slit so that the rim of the flame is on the axis of tube B. Somewhat below the spot where the axis meets the rim, there is the end of the loop formed in a fine platinum wire, which is held by arm E, F is a hollow prism of 60° refractive angle, filled with carbon disulphide.

Also interested in thermochemistry, Kirchhoff determined that the heat evolved or consumed in of a chemical reaction is given by the difference in heat capacity between products and reactants.

Kirchhoff died in 1887 and was buried in the St Matthäus Kirchhoff Cemetery in Schöneberg, Berlin.

Gilbert (G.N.) Lewis

Massachusetts – born Gilbert *G.N.* Newton Lewis (1875 — 1946) is renowned for his work on chemical thermodynamics, the deduction that a covalent bond is formed from a pair, his electronic definition of an acid as an electron donor, the isolation and study of deuterium and his work on phosphorescence and the triplet state (one in which the quantum number for total spin angular momentum is 1).

Home educated until the age of thirteen when he went to the preparatory school of the University of Nebraska, Lewis began his undergraduate studies in Chemistry at Nebraska but after two years transferred to Harvard, graduating in 1896. After a year teaching at Phillips Academy in Andover, he returned to Harvard where he completed a master's degree and then his doctorate and remained at Harvard for another year as a laboratory instructor before moving to Germany to work under Wilhelm Ostwald and Walther Nernst before, once more, returning to Harvard for another three years followed by a year in the Philippines as the country's superintendent of weights and measures. In 1905, he joined the chemistry faculty of the Massachusetts Institute of Technology (MIT) in Cambridge. In 1912, he was appointed dean and chair of the department of chemistry at Berkeley, where he remained until his death at age 70.

Lewi's major field of study was chemical thermodynamics, in which he developed the work of J. Willard Gibbs III who had conceived of the entropy and enthalpy combining to form the free energy of a substance. Lewis spent many years reconciling Gibbs' theory of free energy with experimental measurements of the enthalpy changes in a reaction and a number of empirical laws dealing with the behaviour of ideal gases and dilute solutions. To accomplish this Lewis measured the missing free energy values by either determining the enthalpy or entropy of a substance and the changes of these values occurring in a reaction.

Some of his most famous work, though, and the most relevant to this series is his speculations on the role of the outer electron in chemical bonding and, in his 1916 paper, '*The Atom and the Molecule*', he published his model of the, now classical, covalent bond, as created by the sharing of two electrons by a pair of atoms, with each atom contributing one of the electron pair.

He also introduced the famous dot and cross diagrams to describe a covalent bond and explain the bonding as demonstrated for the bonding in the methane molecule (below) and which are normally known as Lewis structures. In these diagrams, a dot and a cross represent an electron pair with the dot indicating an electron from one atom whilst the cross represents that from the other bonding atom.

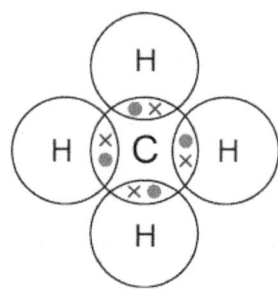

War service in military service meant that Lewis did not return to the theory of covalent bonding until, in 1923, he published his, now famous, monograph '*Valence and the Structure of Atoms and Molecules*' which he published after his theory was promoted and elaborated upon by Irving Langmuir, who between 1919 and 1921, introduced many of the terms commonly used in discussions of bonding such as *covalent* and the *octet rule*.

In the 1920s and 1930s, Lewis', electron – pair, bonding model became crucial in elaborating the mechanisms of many organic reactions and in explaining the formation of coordination and organometallic complexes. He wrote little more about the subject and the concepts were further developed by Linus Pauling who incorporated it into his valence bond theory which was the subject of his classic, 1939, text, '*The Nature of the Chemical Bond*'.

Lewis had by then long turned his attention to the study of deuterium, ^2H, and its compounds and between 1932 and 1934 published twenty four papers on their nature and separation and he was the first to produce, isolate and name heavy water, ^2H_2O also written as D_2O.

He is, however, most famous for his new definition of acids and bases: an acid is an electron donor whilst a base is an electron acceptor which he first proposed, almost casually, in his 1923 monograph. This concept is easily demonstrated by the formation of hydrochloric acid from the dissolution and dissociation of the hydrogen chloride molecule which leads to the formation of the H^+ and Cl^- aqueous ions where the hydrogen atom has lost an electron and hence is an acid whilst chlorine has acted as a base by accepting that electron.

This extended the definition of an acid and a base which extended the Brønsted-Lowry definition which was that an acid is a substance that can donate a proton whilst a base is a substance which can accept a proton.

The Lewis and Brønsted-Lowry definitions are complementary and never contradict each other.

Periodically, Lewis published speculative papers on theoretical physics and, in fact, had as a student postulated that if light is a stream of particles then they might exert a pressure on matter in a vacuum and it was actually Lewis who coined the word *photon* to describe a packet of discrete energy in the electromagnetic spectrum. He was also, in 1909, the first American to write on Einstein's Theory of Special Relativity.

Lewis was never awarded a Nobel Prize despite being nominated forty one times. In his time at Harvard, Lewis supervised the work of two hundred and ninety doctoral students including twenty who went on to be awarded the Nobel prize and also had overall supervision of the work which led to the discovery of all the elements with atomic numbers between 93 and 106.

In 1912 Lewis married Mary Hinckley Sheldon, with whom he had a daughter and two sons. His final research was into the properties and chemistry of hydrogen cyanide and he was found dead in his laboratory on March 23rd 1946. There was speculation that he had committed suicide but this appears to have been unfounded as there was no evidence of the poison in his body and the coroner determined the cause of death to heart disease.

Justus Liebig

Darmstadt – born **Justus Freiherr von Liebig** (1803 – 1873) became renowned for his contributions to agricultural and biological chemistry, and was considered the founder of organic chemistry.

As professor of chemistry at the University of Giessen, he created the modern, practical laboratory – oriented teaching methods. Everybody who has studied chemistry is aware of his name through use of the Liebig condenser (below right).

Liebig was not the first to use a glass tube to collect materials when separated by distillation but he was the first to place an outer jacket with flowing water passing from bottom to top to keep the tube cool.

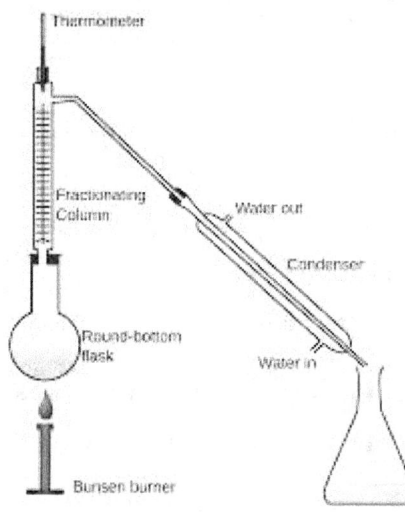

He also invented fractional distillation by placing the condenser vertically above the flask of boiling materials as shown on the left. This was a major advance since complex mixtures of organic compounds could be separated on the basis of their boiling point and forms the foundations of the fractionation of the components of crude oil.

Liebig also determined the essential nutrients for his determination of the need of plants for nitrogen and trace minerals. He devised the *law of the minimum* which stated that the growth of the plant depended on the scarcest nutrient and also devised a method for manufacturing beef extracts as a solid which is still used to produce the world famous '*Oxo*' cube and he is also regarded as the discover of '*Marmite*'.

Liebig also devised a safer method of making the silvered backs of mirrors. Until his work, the process involved mercury which is highly toxic but Liebig discovered that aldehydes reduce silver compounds to metallic silver. A further modification of the method, adding copper to ammoniacal silver salts and sugar, was extremely effective in producing the high quality silver needed for optical mirrors.

Liebig frequently collaborated with Friedrich Wöhler who he had met in 1826 after they had independently reported on the discovery of cyanic acid and fulminic acid. Both compounds had exactly the same chemical composition but very different characteristics and this led to the discovery of chemical *isomers*: molecules of the same chemical composition but different arrangements of the constituent atoms.

In 1832, Liebig and Friedrich Wöhler also published an investigation of the oil of bitter almonds. Through a number of chemical reactions they discovered that a collection of carbon, hydrogen, and oxygen atoms can behave like an atom, take the place of an atom. They had discovered *functional groups*.

In 1826, Liebig, a Lutheran (Protestant), married Henriette '*Jettchen*' Moldenhauer, a Catholic and they resolved their religious difference by determining that their two sons would be brought up as Lutherans (Protestants) and their three daughters as Catholics.

Theodore Lyman

Boston – born **Theodore Lyman** (1874 – 1954) was the scion of a very wealthy family which had arrived in America having travelled from Essex in the 17th century.

After Lyman graduated from Harvard, in 1897, and studying spectra for which he was awarded his doctorate in 1900, he studied with J.J. Thomson in Cambridge.

During the First World War, Lyman served as a military engineer in France with the American Expeditionary Force and achieved the rank of major.

In 1917, Lyman was appointed to a lectureship at Harvard and subsequently a professorship in Mathematics and Natural Philosophy and was also the Director of the Jefferson Physical Laboratory between 1908 and 1917. He only held the professorial position for ten years before resigning, with the title Emeritus Professor, to spend more time on his independent research. He did, however, retain his role at the Jefferson Physical Laboratory for another thirty seven years.

Lyman is renowned for his spectroscopic investigations of the ultra violet region of the electromagnetic spectrum and is famed for his discovery of the, now called, *Lyman Series* which were found to relate to transitions between the lowest energy level i.e. that closest to the nucleus and all other energy levels.

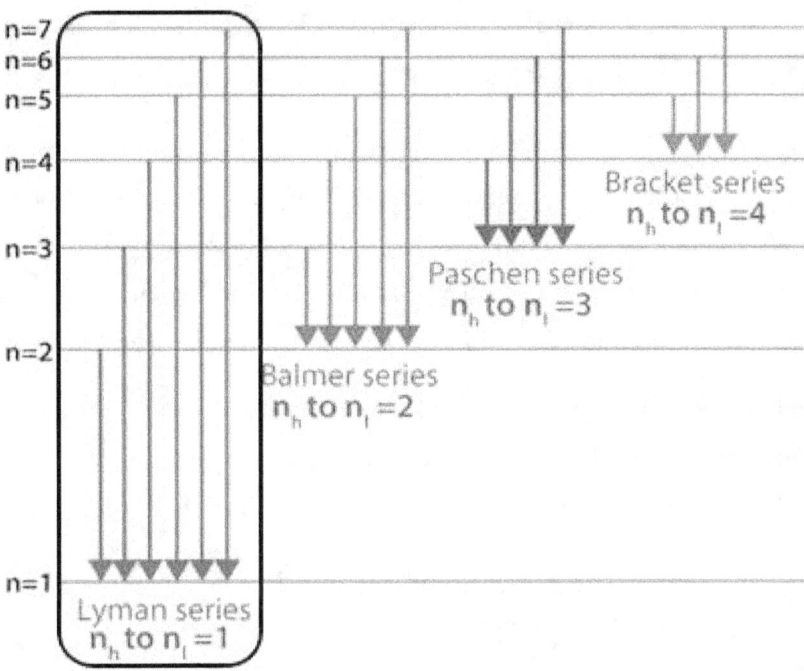

Subsequently, Lyman researched the ultraviolet spectrum further and the use of diffraction of gratings. He was inspired by the work of Victor Schumann (1841 – 1913) who had in, 1893, discovered the ultraviolet region of the electromagnetic spectrum.

Ernest Marsden

The, Manchester – born physicist, Sir **Ernest Marsden** MC FRS (1889 – 1970) was a dual English – New Zealand citizen who is renowned for his work under Lord (Ernest) Rutherford which led to the discovery of the existence of the nucleus.

Marsden studied at the University of Manchester where he worked, as an undergraduate, with Hans Geiger on the Rutherford – inspired *'gold foil experiment'* which determined the existence of a positively charged, dense centre of the atom, the *nucleus*. The apparatus used in the *'gold foil experiment'* was an early version of what became the Geiger – Muller counter which is used to this day to detect α and β emissions.

The Geiger-Muller counter comprises a chamber containing a low pressure gas mixture and two electrodes across which there is a potential difference of several hundred volts. The walls of the tube are either metal or are made of coated metal while the anode is an, axially - mounted wire in the centre of the chamber. When ionising radiation strikes the tube, some molecules of the gas are ionised, creating positively charged ions and free electrons, which are known as ion pairs. The electric field accelerates the positive ions towards the cathode and the electrons towards the anode. This generates a measurable current.

There are further methods to amplify the measured signal whilst the window of the tube can be adapted to make the tube capable of measuring α, β, γ radiation as well as x – rays and, in some circumstances, even neutrons.

In 1915, Marsden, on Rutherford's recommendation, was appointed professor of physics at the Victoria University College in Wellington, New Zealand. This was only temporary since, due to the outbreak of the First World War, Marsden returned to serve in the British Army and spent the years in France as a Royal Engineer on sound – ranging equipment, for which work he was awarded the Military Cross (MC).

After the Armistice and his discharge from the army, Marsden returned to New Zealand and, in 1922, he was appointed Assistant Director of Education and, in 1926, Secretary of New Zealand's Department of Scientific and Industrial Research (DSIR).

In these roles, Marsden initiated a series of projects which kept New Zealand in touch with international developments in the field of radiation and nuclear science. Most importantly, in 1939, he pioneered the non-medical use of radioisotopes.

When World War II broke out, Marsden marshalled New Zealand's scientific work force and managed projects which worked on the use of radar in the Pacific Ocean. He also co-ordinated a team of young scientists who participated in the, American, *Manhattan Project* which resulted in the two nuclear bomb attacks on Japan.

Despite his retirement in 1954. Marsden, now living in Wellington, served on numerous committees, investigated environmental radioactivity having been yet another scientist who had been involved in the *Manhattan Project* to campaign against nuclear weapons.

James Clerk Maxwell

Edinburgh – born **James Clerk Maxwell** FRS FRSE (1831 – 1879) acquired lasting fame for his mathematical combination of electric and magnetic fields to explain the existence of light as electromagnetic radiation.

In his '*A Dynamical Theory of the Electromagnetic Field*' Maxwell demonstrated that electric and magnetic fields travel through space as waves moving at the speed of light By unifying electric and magnetic fields he explained the existence of light, predicted the existence of radio waves and he is regarded as the founder of the discipline of electrical engineering.

Maxwell had already developed the Maxwell–Boltzmann distribution, the statistical description of the kinetic theory of gases and also produced the first colour photograph.

Born in Edinburgh, Maxwell was brought up at Glenlair, a 1,500 acre estate in Kirkcudbrightshire which his parents had inherited. A boy of prodigious talent and curiosity he was educated at home by his mother until her death at the age of eight. He was then taught by his father and his aunt with the added services of the occasional tutor until old enough to attend the Edinburgh Academy when he lodged with another aunt during term time. He excelled in mathematics, English and poetry and published his first scientific paper at the extraordinary age of fourteen on the mathematical description of ellipses in which he developed the ideas of René Descartes. Maxwell completed his studies at the University of Edinburgh where finding his studies undemanding he spent his free time experimenting with home made chemical, electric, and magnetic apparatus. He was, however, most fascinated with the phenomenon of polarised light which he investigated by shining light through gelatine. Through this he discovered the phenomenon of photoelasticity, the means means of determining stresses within physical structures. He published other scientific papers on the elasticity of solids, viscous liquids and oval curves before studying mathematics at Trinity College, Cambridge where he developed his Christian faith and contemplated its relationship to scientific discoveries.

Constantly fascinated by colour, he wrote papers demonstrating that white light could result from a mixture of red, green, and blue light and his, 1855, paper '*Experiments on Colour*' established the principles of colour combination.

Elected a fellow of Trinity at the young age of twenty four, Maxwell was appointed Professor of Mathematics at the University of Aberdeen aged twenty five where he devoted himself not just to investigations but also to teaching and giving free lectures at a local working mens' college.

He also began working on an explanation for the existence of Saturn's rings and established that solid rings would be unstable, fluid rings would break up and concluded that the rings could only be constituted of numerous individual particles which he termed 'brick bats'. His conclusions had to wait verification until the 1980s when they were confirmed during flybys by the Voyager space ship.

In 1860, he was appointed to the Chair of Natural Philosophy at King's College, London and it was there that he made his most important discoveries, produced the first colour-fast photograph, developed the concept of dimensional analysis and became an associate of Michael Faraday.

In London, Maxwell developed his theory of electromagnetic induction, concluding in his 1861 paper, 'On Physical Lines of Force' that magnetic material contained numerous cells of tiny spinning magnets. He also wrote papers on the nature of electrostatics and explained the phenomenon of the plane polarisation of light and its rotation in a magnetic field. Returning to his family estate, Glenlair, in 1865, Maxwell wrote papers on the mathematical behaviour of governors which controlled the speed of steam engines and wrote two books which became standard texts, 'Theory of Heat' (1871) and 'Matter and Motion' (1876).

In 1871, Maxwell became the first Cavendish Professor of Physics at Cambridge and established and developed the Cavendish Laboratory. He remained there until his death from abdominal cancer in 1879 the condition which had killed his mother forty years earlier.

Maxwell's most lasting achievement was, of course, his theory explaining the formation of light through the combination of electric and magnetic fields. He had first commented on fields in 1855 when he read his paper 'On Faraday's lines of force' to the Cambridge Philosophical Society. In this work, Maxwell simplified Faraday's theories and reduced all known knowledge of electricity and magnetism to a series of twenty differential equations with twenty variables. Incredibly, it was actually during a lecture to undergraduates that Maxwell determined the speed of propagation of an electromagnetic field to be the same as the speed of light and came to the conclusion that this could not possibly be a coincidence hence light must be a combination of electric and magnetic fields. He consolidated these ideas in his 1864 paper, 'A Dynamical Theory of the Electromagnetic Field' writing, 'The agreement of the results seems to show that light and magnetism are affections of the same substance, and that light is an electromagnetic disturbance propagated through the field according to electromagnetic laws.' and the twenty equations, later condensed to four partial differential equations by Oliver Heaviside appeared in Maxwell's, 1873, 2 – volume textbook 'A Treatise on Electricity and Magnetism'.

At his time of writing, Maxwell believed that the propagation of electromagnetic waves i.e. light required a medium to carry them and introduced the concept of the luminiferous æther or ether. The existence of the ether was disproved by the famous Michelson and Morley experiment which demonstrated that the velocity of light was the same in all directions which would not be the case if it travelled in a medium of any sort. Moreover, this also meant that it required an absolute frame of reference in which the equations were valid meaning that the equations changed form for a moving observer. Since equations should be independent of the observer this inspired Albert Einstein when formulating his theory of special relativity and enabled Einstein to dispense with the concept of the æther.

Although Maxwell's discoveries of the relationship between electric fields, magnetic fields and light are his overwhelming achievement, his work on colours and colour photography and on kinetic theory of gases is also highly significant. Since all colours could be produced by the combination of three different colours, Maxwell reasoned that if three black and white photographs were taken through red green and blue filters and the three transparent images were projected onto the same space then the photograph should appear coloured and he demonstrated this in an 1861 lecture at the Royal Institution. Maxwell developed a theory of the velocity of gas particles which was adopted and later generalised as the Maxwell-Boltzmann distribution.

A man of widespread interests and extraordinary achievements, Maxwell's early death was a loss to all of us.

Dmitri Mendeleev

Siberian – born **Dmitri Ivanovich Mendeleev** (1834 – 1907) is renowned for his development of the Periodic Table.

Mendeleev was far from the first to attempt to determine patterns in the physical and chemical properties elements but he is renowned forever for formulating his *Periodic Law* and leaving gaps where no elements fitted in. In this way he predicted the existence of eight, previously, unknown elements.

The most famous is his conclusion that there should be an element with properties similar to *boron*. He termed it *eka-boron* and we know it as *scandium* and amongst the other predicted elements he predicted the properties of *eka – silicon* which is now called *germanium* and is an essential element in semiconductors.

Mendeleev was the youngest of seventeen children and his father was a teacher who lost his position when he went blind forcing his mother to re-open her family's abandoned glass factory. This lasted until, aged thirteen, his father died and the glass factory was destroyed by fire. After his secondary education in Tobolsk, Mendeleev was taken by his mother to Moscow where he was rejected by the university and then to St. Petersburg which university did accept him.

Mendeleev moved to the Crimea to recuperate from tuberculosis and, having done so, returned to St. Petersburg in 1857, working as a lecturer and publishing a textbook on organic chemistry. In 1867, having secured tenure and a professorship, Mendeleev turned to inorganic chemistry and hence to his formulation of the Periodic Table.

In 1867, Mendeleev wrote his definitive textbook, *Principles of Chemistry* which was published in two volumes, in 1868 and 1870 and contained his theorisation of the table of the elements and the famous gaps, predicting the existence of previously undetected elements.

Mendeleev also investigated the composition of petroleum and was instrumental in the building of the first oil refinery in Russia and predicted the uses of many of the fractions of crude oil.

There is a myth that he was responsible of the formulation of vodka as containing, by volume, 40% ethanol. This is nonsense since the percentage by volume had been established when Mendeleev was nine years old but was nurtured by the knowledge that his doctoral thesis had been concerned with the properties of varying solutions of ethanol in water at concentrations of ethanol above 70% (v/v) but did not mention vodka.

Mendeleev is commemorated for all time by the naming of element of atomic number 101 after him: mendelevium, $_{101}$Md which is an artificial element created by the bombardment of Einsteinium, $_{99}$Es, with alpha particles.

Henry Moseley

Weymouth – born **Henry** 'Harry' **Gwyn Jeffreys Moseley** (1887 – 1915) made one of the most fundamental and important contributions to atomic physics when he identified the atomic number, the number of protons in the nucleus, as *the* fundamental property of an element.

His atomic number theory refined Rutherford's concept of the structure of the atom and provided support for Bohr's theory of the electronic distribution of the electrons.

Graduating from Oxford in 1910, Moseley worked with Lord (Ernest) Rutherford until offered a position at Oxford. In 1912, whilst experimenting with the energy of beta particles, Moseley showed that high potentials were attainable from a radioactive source of radium, conceiving the idea of atomic energy.

The following year he observed and measured the x-ray spectra of many metals and, using Bragg's law to determine the wavelengths of the x-rays he discovered a systematic mathematical relationship between the wavelengths of the diffracted waves and the atomic numbers of the targeted metals. This relationship became known as *Moseley's law*.

Before Moseley's discovery, the atomic numbers were regarded as less important than *atomic weights* (as relative atomic masses were then called). Mendeleev had ignored the atomic weights of a number of pairs of metals, notably cobalt and nickel, when assembling his Periodic Table. Moseley's discovery demonstrated that Mendeleev had been correct to do this when the same table was ordered using the atomic number.

In addition, Moseley showed that there were missing elements in the periodic Table as he could find no spectra and, hence no elements, corresponding to atomic numbers 43, 61, 72, and 75. He, therefore, predicted the existence of these elements which are now known and are called *technetium*, *promethium*, *hafnium* and *rhenium* respectively.

It is impossible to overestimate the contribution that Moseley made to atomic physics as his work supported Rutherford's concept. In the words of Niels Bohr (1962) '*Originally Rutherford's work was not taken seriously at all and that the great change came from Moseley.*' whilst Robert Millikan wrote, '*In a research which is destined to rank as one of the dozen most brilliant in conception, skilful in execution, and illuminating in results in the history of science, a young man twenty-six years old threw open the windows through which we can glimpse the sub-atomic world with a definiteness and certainty never dreamed of before. Had the European War had no other result than the snuffing out of this young life, that alone would make it one of the most hideous and most irreparable crimes in history.*'

On the outbreak of the First World War, Moseley left Oxford to volunteer for the Royal Engineers with whom he served as a communications officer. Tragically, Moseley was killed at Gallipoli when he was shot in the head.

Isaac Newton

Sir **Isaac Newton** PRS (1642 – 1727) is renowned for his law of gravity and his equations of motion, discovery of white light being composed of the colours of the rainbow and these achievements are so well known that there is no need to elaborate on them here. There was, however, much more to Newton as he was a noted mathematician, theologian and author.

In his classic, 1687, book, *Philosophiæ Naturalis Principia Mathematica* (*Mathematical Principles of Natural Philosophy*) Newton laid out the principles of classical mechanics which remain correct to this day for particles but which has been in some instances has been superseded by Einstein's theories of relativity.

Equally importantly, he also devised the infinitesimal calculus where the gradient of the slope of a curve is known as the differential and the area underneath a curve is the integral and he demonstrated how closely they are related. In later years, his notation was abandoned in favour of the simpler symbolism defined at almost the same time by Gottfried Wilhelm Leibniz, whose notation $\delta y/\delta x$ for a gradient is used to this day.

Newton eradicated any doubt that the Earth orbits the sun, explained that the twice – daily sea tides were caused by the moon's orbit of the Earth and predicted that the Earth is not a perfect sphere but is oblated i.e. flattened at both poles.

Newton also built the first reflecting telescope and in his, 1704, work '*Opticks*' developed his theory of the light based on the separation of white light into the colours of the rainbow using a prism.

Newton also deduced the law of cooling which states that the hotter an object the more quickly it will cool down, made the first theoretical calculation of the speed of sound, reverting back to mathematics generalised the binomial theorem to non-integer exponents, developed a method for approximating the roots of a function and also developed his interpolation process for calculating the value of an unknown quantity present within a series or on a curve.

A fellow of Trinity College, Cambridge and the second Lucasian Professor of Mathematics, Newton was a devout Christian but one who privately rejected the concept of the Holy Trinity (Father, Son and Holy Spirit), believing in the existence of a Unitarian God as a single deity.

He also investigated alchemy and there have been suggestions that Newton invented the story of the apple falling from the tree to prevent suggestions that he was involved in magic which at the time was a capital offence. In any event, Newton developed his theory of gravity whilst at his family home, Woolsthorpe Manor, in Lincolnshire to escape the latest outbreak of the bubonic plague and it was also there that he conducted much of his work on light.

Newton was, politically, a Whig and served two terms as Member of Parliament for the University of Cambridge (1689–90, 1701–02) and was also Master of the Royal Mint for almost thirty years.

Hans Christian Ørsted

Rudkøbing – born **Hans Christian Ørsted** (1777 – 1851) made the hugely significant discovery that an electric current generated a magnetic field. He discovered this by placing a tiny magnet alongside the current – carrying wire and noted the deflection of the needle of the compass. He was thus the first person to identify any relationship between electric fields and magnetic fields which laid the basis for the theory of light proposed by James Clerk Maxwell.

Ørsted also demonstrated that two wires carrying an electric current in the same direction would attract each other and conversely two wires carrying current in opposite directions would repel each other.

This discovery alone had enormous consequences since it led to *Hund's Rule* which stated that no orbital can contain more than two electrons.

The son of an apothecary, a pharmacist in modern terminology, Ørsted displayed an early interest in chemistry. Home – educated and also self – taught, Ørsted studied at the University of Copenhagen, where he and his brother studied physics and philosophy. In those times the discipline was known as *Natural Philosophy* and, in 1799, Ørsted was awarded his doctorate for a thesis on the works of the philosopher Emmanuel Kant.

In 1800, Ørsted read about the invention by Alessandro Volta of the voltaic pile (the first constructed electrical battery) and a year later he received a travel scholarship and a grant which enabled him to spend three years travelling across Europe visiting scientific laboratories, notably those in Berlin and in Paris to investigate electricity.

In Berlin, Ørsted met Johann Wilhelm Ritter who believed there was a connection between electricity and magnetism which reconciled with Ørsted's concept of the unity of nature. In 1806, Ørsted became a professor of physics at the University of Copenhagen where he conducted his research in both electric currents and acoustics.

In 1825, Ørsted also made a most significant contribution to chemistry, technology and industry by isolating metallic aluminium by reduction of the white powder, aluminium chloride.

Aluminium is best known for its uses in drink cans etc; but being light (low density) yet strong has uses in aeroplane wings and aircraft fuselages.

Louis Paschen

Schwerin – born Louis Carl Heinrich Friedrich Paschen (1865 – 1947) is known for his work on electrical discharge tubes and for the discovery of his eponymous series of infrared hydrogen spectral lines which he observed in 1908.

After completing his studies at the universities of Berlin and Strasburg, Paschen worked at the Academy of Münster before being appointed a professor of physics at the Technical Academy of Hanover (1893) and then at the University of Tübingen (1901) before moving to the University of Berlin where he taught until his death. Paschen became renowned as the most talented spectroscopist of his generation and he also investigated Kirchhoff's function which related temperature to wavelength as well as determining the infrared absorption of carbon dioxide and water vapour. Writing this at the time of the Covid19 crisis (June 2020), his work has become of critical use and is the basis of the infrared detector pointed at the forehead to determine one's temperature and it is also of great use in heat seeking equipment used by by the police and by the military forces.

In 1895, working with Carl Runge and excited by Ramsay's announcement of the discovery of previously unidentified lines, Paschen determined that the spectral lines matched some of those observed in sunlight and hence demonstrated the existence of helium on both Earth and in the sun. The Paschen Series as shown below

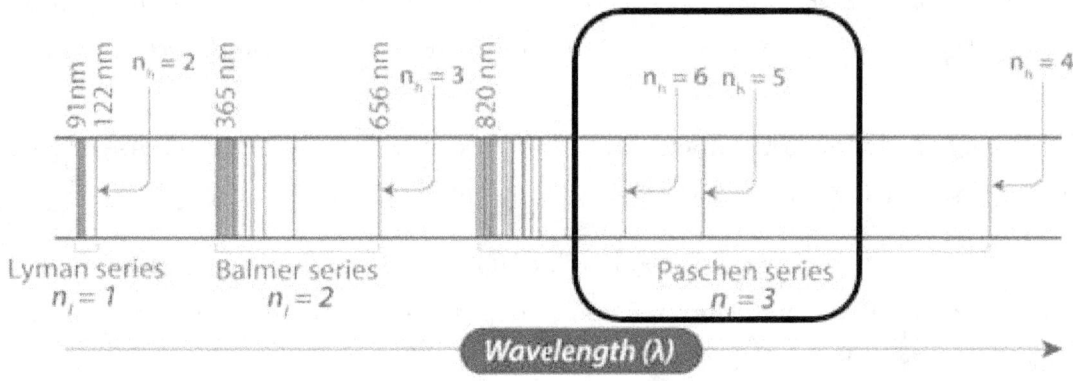

was shown to relate to transitions between the third energy level and higher levels as depicted below as the circular orbits (left) and the horizontal lines (right).

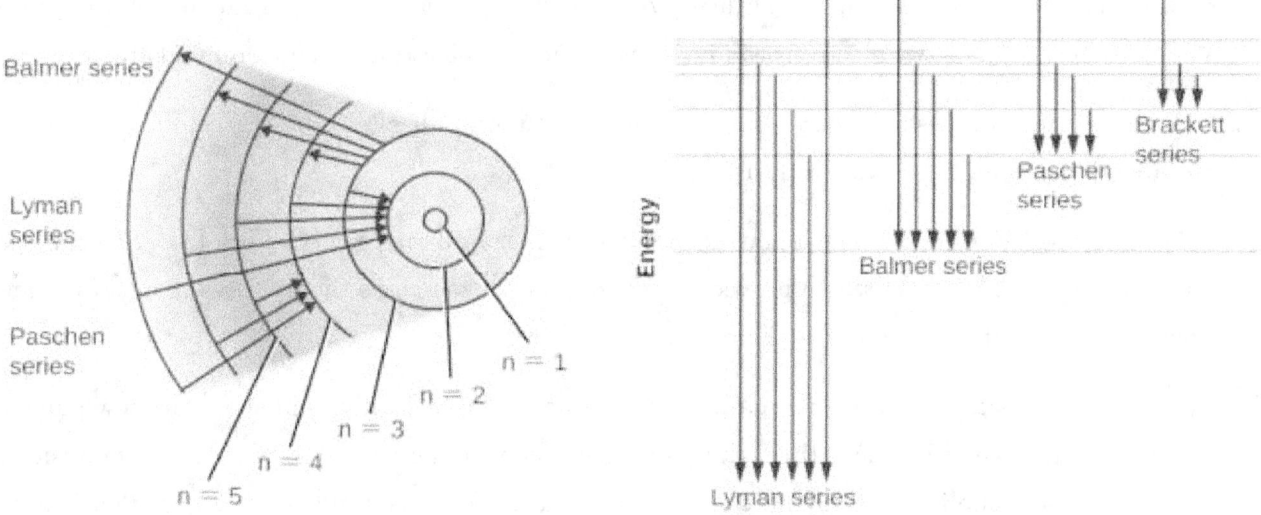

Wolfgang Pauli

 Vienna - born Wolfgang Ernst Pauli (1900 – 1958) was a precocious theoretical physicist who was awarded his doctorate at the age of twenty one having only left school aged eighteen and received the 1945 Nobel Prize in Physics *'for the discovery of the Exclusion Principle, also called the Pauli Principle.'*

Pauli studied under Arnold Sommerfeld who had determined the significance of two quantum numbers, the *azimuthal quantum number* which describes the shape of orbitals and the *magnetic quantum number* which identifies the number of mutually degenerate orbitals such as the three p - orbitals present in all energy levels greater than the first energy level.

There are four such quantum numbers, the first is the *principal quantum number* which describes the energy levels, the second is the *azimuthal quantum number*, the third is the *magnetic quantum number* and the fourth is the *spin quantum number*.

In chemistry we visualise these as follows:

- *Principal quantum number*: energy level
- *Azimuthal quantum number*: shape of orbitals e.g. s, p, d and f
- *Magnetic quantum number*: number of orbitals of each type i.e. there is always one s - orbital in each energy level, from the second energy level there are always p - orbitals and the third also contains d-orbitals etc;
- *Spin quantum number*: the orientation of the electron which can spin in a clockwise or anticlockwise direction. These spins are denoted as +/- ½.

Sommerfeld also asked Pauli to review the theory of relativity for the *Encyclopedia of Mathematical Sciences* and his 237 – page monograph was highly praised and recommended by Einstein. To this day, it is one of the standard introductory texts on Special and General relativity.

Pauli subsequently spent a year as an assistant to Max Born at the University of Göttingen before moving to Copenhage to work with Niels Bohr before becoming a lecturer at the University of Hamburg in 1923 where he remained for five years and formulated the *Pauli Exclusion Principle* which, in chemical terms states that in an atom or molecule, no two electrons can have the same four electronic quantum numbers.

Since there are only two possible values for the spin quantum number (+ ½ or - ½) and this means that no orbital can contain more than two electrons.

In 1926, Pauli used Heisenberg's new matrix theory of modern quantum mechanics to derive the observed spectrum of the hydrogen atom which proved the validity of Heisenberg's theory. This result was important in securing credibility for Heisenberg's theory.

Despite his numerous contributions to quantum mechanics, Pauli actually published very few papers as he much preferred discussing his ideas through verbal discussions and correspondence with key physicists who copied his letters and publicised his ideas, concepts and conclusions essentially as they were clearly extremely important.

The importance of this correspondence is demonstrated by a 1930 letter to Lise Meitner* and others in which he demonstrated a wry sense of humour. He had considered the beta decay of natural radioactive elements and commenced the letter *"Dear radioactive ladies and gentlemen"*.

This letter was stunningly important since he proposed the existence of another neutral particle but with less than 1% of the mass of a proton whose existence explained certain phenomena observed in beta decay. This was just before Chadwick's discovery of the neutron and, in 1934, Enrico Fermi termed the particle the *neutrino*, hence coining a new Italian word meaning *'little neutral one'*. The existence of the neutrino was confirmed experimentally in 1956.

The, 1938, annexation of Austria made Pauli a German citizen, which he resisted, and he moved to the United States in 1940 where he became professor of theoretical physics at the Institute for Advanced Study at Princeton and he became a naturalised American citizen in 1946 before moving to Zurich in 1949 where he also gained naturalised Swiss citizenship and where he remained until his premature death from pancreatic cancer.

Pauli was also fascinated by parapsychology and corresponded regularly with Carl Jung and, in 1955, they published a seminal book on the subject, *'The Interpretation of nature and the psyche Synchronicity: an acausal connecting principle'*.

He also delighted in the affectionately terms *'Pauli effect'* which was named after it was noticed that, very often, experimental apparatus broke, merely apparently because of his presence. This resonates with the reputation of J.J. Thomson who was apparently notoriously clumsy. There are examples of Thomson's apparatus in museums but the joke went that he never touched them because since they still worked and were unbroken they must be replicas.

In 1929, Pauli married a cabaret dancer, Käthe Margarethe Deppner but the marriage ended in divorce within a year and he married Franziska Bertram in 1934. He died in 1958 leaving no descendants.

*As this monograph is not about radioactivity there is unfortunately no space to record the life and achievements of one of the pioneers of radioactivity experimentation, Lise Meitner, who was fundamental to the discovery of uranium fission and hence nuclear weapons and power.

Never awarded the Nobel Prize, Meitner (pictured right) holds the record for the number of nominations which comprises an extraordinary 19 nominations for the Nobel Prize in Chemistry (between 1924 and 1947) and an even more incredible 29 times for Nobel Prize in Physics (between 1937 and 1965).

In acknowledgement of the the undeserved slight her achievements were recognised by an invitation to attend the exclusive Lindau Nobel Laureate Meeting in 1962.

Linus Pauling

There have been scientists, who have been awarded the Nobel Prize more than once in a scientific discipline, such as John Bardeen, Nobel Laureate in Physics (1956 and 1972), Frederick Sanger, Nobel Laureate in Chemistry (1958 and 1980) and Marie Curie, Nobel Laureate in Physics (1904) and Nobel Laureate in Chemistry (1911). There has, however, been only one person to become a Nobel laureate in two entirely unrelated disciplines and that is Linus Pauling (Nobel Prize in Chemistry, 1954, and for Peace, 1962).

Pauling (1901 – 1994) was one of the most significant founders of both the fields of quantum chemistry and molecular biology. His contributions to the theory of the chemical bond included the concept of orbital hybridisation and the most precise scale of the electronegativities of the elements. He also worked on the structures of biological molecules, demonstrated the importance of the alpha helix and beta sheet in protein secondary structure and inspired the work of Watson, Crick, Wilkins and Frankland in their determination of the double helix structure of the DNA molecule. From his work on the alpha helix, Pauling decided that DNA was constructed of a triple helix. If he had not missed out the double helix structure it is highly likely that he would have become the first, and probably now only, person to be awarded three Nobel Prizes.

Pauling was educated at Oregon Agricultural College and, having saved enough money from part time jobs he was able to fund his studies at Oregon State University from which he graduated with a degree in chemical engineering and it was here that he met his future wife, Ava Miller. Upon graduation, he went on to postgraduate study at California Institute of Technology in Pasadena (Caltech) where he studied the application of x-ray crystallography on crystalline minerals, receiving his doctorate in physical chemistry and mathematical physics in 1925.

In 1926, Pauling, awarded a Guggenheim Fellowship to travel to Europe, went to study under Arnold Sommerfeld in Munich, Niels Bohr in Copenhagen and Erwin Schrödinger in Zürich since Pauling had become intrigued in the application of quantum mechanics to chemical bonding and which became the subject of his future research. Appointed to the position of assistant professor at Caltech he concentrated on x-ray crystallography and the quantum mechanical calculations on atoms and molecules, publishing fifty papers in five years. Appointed a full professor in 1930, he published, two years later, the paper which really launched his career when he proposed that one s – and three p – orbitals, which overlap, could hybridise themselves into four chemically, magnetically and mutually degenerate orbitals, which he designated as sp^3 – orbitals and hence was able to explain the bonding in the methane, CH_4, molecule which contains four $C – H$ bonds of equal strength as shown by their equal lengths. This is discussed further in the glossary in the next section.

In the same year, Pauling also introduced his scale of electronegativities. This was not a new concept but Pauling was the first to quantify the values on the basis of bond energies and his scale in the only one now used. Electronegativity is discussed earlier in this volume.

In 1937, Pauling was appointed to be the George Fischer Baker Non-Resident Lecturer in Chemistry at Cornell University where he delivered nineteen lectures and wrote what has been described as the most influential chemistry textbook of the 20th century, '*The Nature of the Chemical Bond*' and it is still cited as reference more than eighty years after its first edition appeared.

In this book, Pauling also explained the nature of the bonding in the benzene molecule and explained that the Kekulé structure of benzene which predicted three single C − C bonds alternating with three C = C bonds which does not occur could be explained by the presence of a molecular orbital existing both above and below the six − carbon ring and is discussed earlier in this volume.

Becoming ever more interested in molecular biology, Pauling showed that enzymes bring about reactions by stabilizing the transition state of the reaction, that DNA reproduction is due to complementarity between the functional groups and not due to similarity and he was also one of the authors to first identify that sickle cell anaemia is caused by the presence of a mutated protein which does not prevent the formation of normal forms of haemoglobin but also creates mutated forms of the molecule.

Until the advent of the Second World War, Pauling had been apolitical but declined an invitation to work on the Manhattan project to create a nuclear weapon but did perform military research devising an instrument for measuring the oxygen conditions in aircraft and submarines. Subsequently, this unpatented invention was developed, commercially, for use in incubators for premature babies. He also worked on explosives, rocket propellants and armour − piercing shells and, in 1948, he was awarded the Presidential Medal for Merit by President Harry S. Truman. In the aftermath of the nuclear bombing of Hiroshima and Nagasaki, and supported by his wife, Pauling became a peace activist and campaigned actively against the proliferation of nuclear weapons. In 1946, he joined the Emergency Committee of Atomic Scientists, chaired by Albert Einstein, whose mission was to warn the public of the dangers associated with the development of nuclear weapons. This activism led at the height of McCarthyism to the State department denying him a passport which was only returned in full to enable him to attend his 1954 Nobel Prize ceremony.

In 1958, Pauling published 'No more war!', in which he called not just for an end to the testing of nuclear weapons but also for an end to war itself. His work in the determination of the concentrations of strontium − 90 in baby teeth led to a moratorium on above-ground nuclear weapons testing which led to the Partial Test Ban Treaty which was signed, in 1963, by John F. Kennedy and Nikita Khrushchev. On the day that the treaty went into force, October 10, 1963, the Nobel Prize Committee awarded Pauling the Nobel Peace Prize for 1962. There was much opposition at Caltech to his political activism and he was forced to step down as Chairman of the Chemistry and Chemical Engineering Division. Although retaining tenure as a full professor, Pauling resigned from Caltech and, in 1969, he became a professor of chemistry at Stanford University, having spent the intervening years on on his peace activism, funded by his Nobel Prize money. Continuing in this vein, Pauling was a vocal opponent of the Vietnam War.

Linus Pauling and Ava Helen Miller were married for fifty eight years until her death (1981) and they had four children all of whom became eminent in the respective fields.

Pauling died of prostate cancer in 1994 and on March 6th, 2008, the United States Postal Service released a 41 cent stamp honouring Pauling, who had in his lifetime published more than 1200 papers and books, created the concept of molecular orbital theory, crystallised the theory of covalent bonding and worked strongly for peace.

August Pfund

Wisconsin – born **August Herman Pfund** (1879 – 1949) studied at the University of Wisconsin–Madison and after graduation in 1903 moved to Johns Hopkins University in 1903 where he received his doctorate and remained for his entire career eventually becoming a full professor and head of the physics department.

Pfund discovered the fifth series in the hydrogen spectrum i.e. the series where an electron is promoted from the fifth level, absorbing radiation, or an excited electron relaxes and returns down to the fifth level emitting exactly the same wavelength required to have promoted it in the first place.

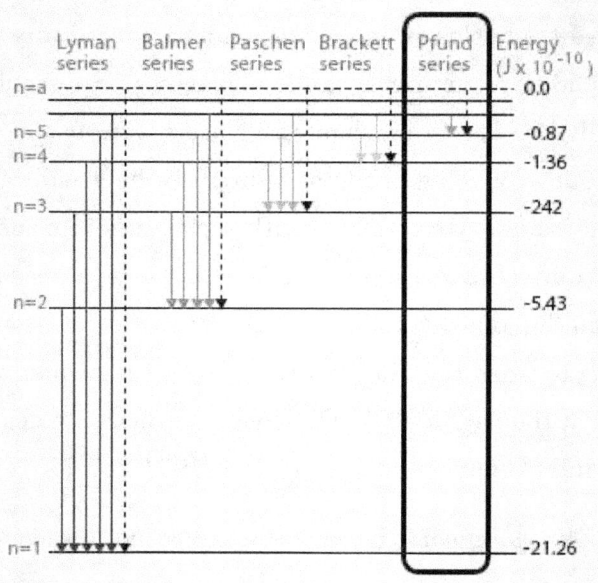

which can also be visualised as orbits:

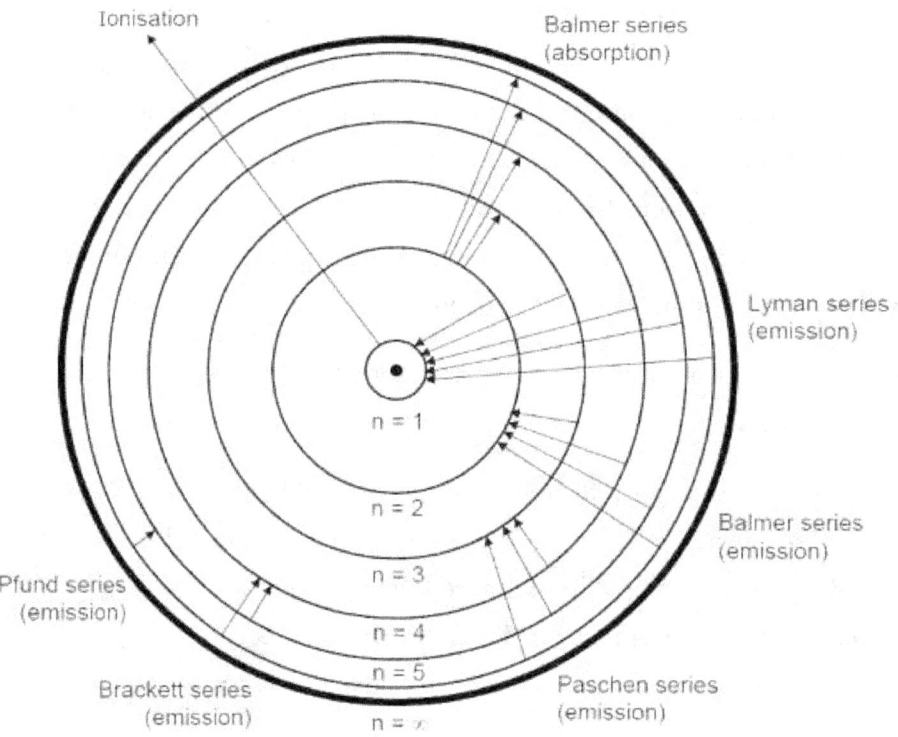

Max Planck

Kiel – born **Max Karl Ernst Ludwig Planck** (1858 – 1947) was a mathematician and theoretical physicist whose concept of the quantisation of energy earned him the 1918 Nobel Prize in Physics.

Until Planck's work, light energy had been considered to be a continuum i.e. particles could have any quantity of energy. Planck decided that energy must exist in small packets to explain a mathematical conundrum and did not, initially, consider it physically significant. He made this assumption when studying black body radiation. Several theories had worked at high frequencies but completely failed at low frequencies. Planck realised that one solution to the problem, which explained the behaviour of a black body, was to assume that electromagnetic radiation comprised a series of packets of energy whose energy was proportional to its frequency and, in 1900, presented his theory to the German Physical Society including the equation

$$E = h\nu$$

where E is energy, ν is the frequency and h is a proportionality constant 6.634×10^{-34} Js^{-1}.

The great grandson and grandson of eminent theology professors in Göttingen and with a father who was a law professor at the Universities of Kiel and Munich, his family moved to Munich in 1867 where he excelled at mathematics and physics but was also found to be exceedingly musically gifted; a talented singer, pianist, organist, cellist and composer.

Planck's father's friend, the physicist Philipp von Jolly attempted to persuade Planck to study music not physics stating, with regard to physics *'In this field, almost everything that can be discovered is already discovered, and all that remains is to fill a few holes.'*. Planck insisted that he was interested in understanding the fundamentals of the discipline and had no desire to discover new things and started studying physics at the University of Munich in 1874 concentrating on theoretical physics. He was not an experimental physicist and had little interest in becoming one.

In 1877, Planck went to the Friedrich Wilhelms University in Berlin to study under Hermann von Helmholtz, Gustav Kirchhoff and, the mathematician, Karl Weierstrass and was awarded his doctorate in 1880 for his work in thermodynamics. Planck spent most of the remainder of the 19th century lecturing in Berlin. He spent some time at Columbia University (New York City) before returning to Berlin where he remained until his retirement in 1926 when he was succeeded by Erwin Schrödinger.

Max Planck was married twice. With his first wife, Marie Merck, he had four children, Karl (1888–1916), twins Emma (1889–1919) and Grete (1889–1917) and Erwin (1893–1945). Marie died of tuberculosis in 1909 and in 1911, Planck married Marga von Hoesslin (1882–1948) with whom he had a fifth child, Herman.

During the First World War, Erwin was taken prisoner by the French in 1914 and spent the war years as a prisoner, Karl was killed in action at Verdun whilst Emma and Grete both died in childbirth.

In January 1945, Erwin, to whom Planck had been particularly close was murdered by the Nazis after his involvement in the 1944 plot to assassinate Hitler had been discovered.

Edward Purcell

Illinois – born **Edward Mills Purcell** (1912 – 1997) shared the 1952 Nobel Prize in Physics for his, 1946, discovery of nuclear magnetic resonance in liquids and in solids. His co-recipient was Felix Bloch.

Purcell studied electrical engineering at Purdue University and then pursued post – graduate research at Harvard. He spent the war years working on the development of microwave radar at MIT.

In December 1946, working with Robert Pound and Henry Torrey, he discovered nuclear magnetic resonance (NMR) which is discussed earlier in this volume.

Also researching in astronomy, Purcell was the first to detect radiowave emissions from galactic hydrogen emissions which led to the first views of the spiral arms of the Milky Way and hence was fundamental in the foundation of radio astronomy.

He also made significant contributions to solid state physics, making seminal discoveries in the study of solid state physics, with studies of spin-echo relaxation, nuclear magnetic relaxation, and negative spin temperature which was fundamental to the development of the laser) and also studied symmetry in particle physics.

In 1965, he wrote what became the standard textbook '*Electricity and Magnetism*' in which he was the first writer to discuss the significance of relativity to this area. Freely available since it was funded by a government grant, the book was originally published as one of the volumes of the Berkeley Physics Course.

He also made significant to biology when he related the Reynolds Number to plant systems. The Reynolds Number (Re) relates the velocity of the flow of a fluid to its viscosity and the diameter of the tube according to:

$$Re = \frac{uD}{\upsilon}$$

where

u is the flow speed of the fluid, D is the diameter of the tube and υ is the kinematic viscosity of the fluid.

The recipient of many awards, Purcell served as science advisor to Presidents Dwight D. Eisenhower, John F. Kennedy, and Lyndon B. Johnson. He was elected President of the American Physical Society, and was a member of the American Philosophical Society, the National Academy of Sciences, and the American Academy of Arts and Sciences and was awarded the National Medal of Science in 1979.

Isidor Isaac Rabi

Polish – born **Isidor Isaac Rabi** (1898 – 1988) received the 1944 Nobel Prize in Physics for his discovery of nuclear magnetic resonance.

Brought up in New York's Lower East Side, Rabi started studying electrical engineering at Cornell University but quickly switched to chemistry. Being Jewish he was, due to anti – Semitism, unable to obtain employment in the chemical field and instead he undertook postgraduate studies in physics at Columbia University.

Rabi became fascinated by quantum mechanics and went to Europe in 1927 hoping to work with Schrödinger. That was not possible as Schrödinger was leaving to become head of the Theoretical Institute at Friedrich Wilhelm University in Berlin and, instead, Rabi worked at the University of Munich as a post – doctoral student before moving to Copenhagen to study under Niels Bohr and then with Wolfgang Pauli at the University of Hamburg. It was at Hamburg that Rabi met Stern and they collaborated on modifications of the Stern – Gerlach experiment to extend it to other applications. In Leipzig, Rabi met Werner Heisenberg and the leader of the future, Manhattan Project, Robert Oppenheimer with whom he became lifelong friends and the two collaborated on many research projects. Rabi and Oppenheimer both went to ETH Zurich to work with Wolfgang Pauli before Rabi returned to Columbia.

It was at Columbia that he predicted that the Stern – Gerlach experiment, which had demonstrated that atoms had angular momentum commonly known as *spin* but could only assume particular values i.e. quantised, could be modified to investigate the properties of sub-atomic particles. This demonstrated that both all sub-atomic particles possess the property of spin which could only be multiple of +/- ½.

Rabi worked with Victor W. Cohen to build his first molecular beam apparatus which employed a weak magnetic field in an experiment to detect the nuclear spin of sodium from which they concluded that the main isotope of sodium, ^{23}Na, has a nuclear spin of 3/2. Rabi and his co-workers next turned attention to the hydrogen atom of which three isotopes were known, ^{1}H, ^{2}H (deuterium) and ^{3}H (tritium). Deuterium, ^{2}H, had only recently been discovered at Columbia by Harold Urey who received the 1934 Nobel Prize in Physics for this discovery. Urey supplied Rabi's group with samples of gaseous deuterium, ^{2}H$_2$, and heavy water, ^{2}H$_2$O, commonly written as D$_2$ and D$_2$O respectively and supported Rabi to the extent that he donated half of his Nobel Prize money to establish further facilities at Rabi's Molecular Beam Laboratory.

Rabi used nuclear magnetic resonance to measure the magnetic moment and nuclear spin of the nucleus of many different isotopes and it was for this that he became a Nobel Laureate.

At the suggestion of Cornelis 'Cor' Gorter, the team experimented using an oscillating field and this became the basis for the nuclear magnetic resonance experimental method. Rabi spent the war years working on radar at MIT and then participated in the Manhattan Project to develop the atomic bomb. He was instrumental in the establishment of the Brookhaven National Laboratory in 1946 and with the establishment of CERN in 1952.

Rabi died of cancer in 1988 and when his cancer was diagnosed he was examined by magnetic resonance imaging which had arisen from his own research and development.

Ernest Rutherford

The *father of nuclear physics*, **Ernest Rutherford**, 1st Baron Rutherford of Nelson, OM, FRS HFRSE (1871 – 1937), is renowned for the discovery of the atomic nucleus as described and discussed earlier in this volume but his achievements extended well beyond this.

Born in Nelson on the South Island, Rutherford, son of a Scottish father, James, and an English mother, Martha Thompson, studied at Canterbury College, University of New Zealand. A noted rugby player, during these years he invented a radio receiver and became famous for being able to receive radio signals over half a mile, briefly becoming the holder of the world record for the distance of radiowave transmission until outdone by Marconi.

Rutherford performed his postgraduate studies at the Cambridge Cavendish Laboratory working under Sir J.J. Thomson. In the early years of the 20th century Rutherford was a professor at McGill University in Quebec where he discovered the concept of the radioactive half – life, the radioactive element *radon* and differentiated between and named *alpha*,α, and *beta*, β, radiation as well as determining that beta radiation and cathode rays are identical and hence the same.

At McGill he was joined by Frederick Soddy and they discovered, through the study of thorium, the half life of radioactive elements. It was for this work that Soddy was awarded the 1921 Nobel Prize in Chemistry.

In 1903, Rutherford determined that radiation discovered in 1900 by Paul Villard in 1900, as an emission from radium, must be different to his already named, α and β, radiation since it had much greater penetrating power. Given that it was possible that there might be still other types of radiation as yet undetected, Rutherford termed this radiation *gamma*, γ ,radiation leaving space in the alphabet for other types. Other types of radiation have, indeed, been detected but his terminology, α, β and γ, is used to this day for the three most well known types of radiation.

In 1908, Rutherford was awarded the Nobel Prize in Chemistry *'for his investigations into the disintegration of the elements, and the chemistry of radioactive substances'*.

By then, Rutherford had moved to the University of Manchester and it was there that he and Thomas Royds proved that *alpha* radiation has the same composition as a *helium* nucleus – containing two protons and two neutrons.

In 1911, although unable to theoretically establish if it was positively or negatively charged, Rutherford theorised that atoms have the vast majority of their mass concentrated in a very small area in the centre of the atom. This inspired his *'gold foil experiment'* which was conducted by Hans Geiger and Ernest Marsden and is described and discussed in detail earlier in this volume.

During the First World War Rutherford worked on the sonar detection of submarines and in 1917, he performed the first artificially induced nuclear reaction when nitrogen molecules were bombarded with alpha particles and detected the emission of a subatomic particle which, in 1920, he termed the proton.

By then Rutherford had succeeded Sir J.J. Thomson as Director of the Cambridge Cavendish Laboratory and in 1921, collaborating with Niels Bohr, he predicted the existence of the *neutron*, a term he first used in his 1920 Bakerian Lecture. The neutron was predicted on the basis that neutral particles of the same mass of the proton could somehow keep the nucleus together since protons should repel each other, force the disintegration of all nuclei and would also account for the *missing mass* of the nucleus which could not be accounted for by summing the masses of the constituent protons and electrons. He was thus responsible for the naming of both nuclear particles, the *proton* and the *neutron* as well as the three first detected types of radioactive emissions α, β and γ.

Rutherford's theory of neutrons was proved in 1932 by his student James Chadwick in his elegant experiment for which Chadwick received the 1935 Nobel Prize in Physics. Rutherford, therefore, supervised the work of four students, Soddy, Chadwick, Cockcroft and Walton who all earned their Nobel Prizes for experiments conducted under his leadership. Only Sir J.J. Thomson has exceeded that record with eight of his researchers, including Rutherford, progressing to becoming Nobel Laureates.

In 1925, Rutherford pressured the Government of New Zealand to create the Department of Scientific and Industrial Research (DSIR) which was established the following year. In 1925, he was appointed to the Order of Merit and was elected President of the Royal Society in which position he served for five years.

In the 1930s he served as President of the Academic Assistance Council which supported almost 1,000 German academic refugees who had fled Germany following Hitler's rise to power and was raised to the peerage in 1931 as First Baron Rutherford of Nelson, New Zealand, and Cambridge in the County of Cambridgeshire.

Rutherford was also a prolific writer on popular science and his notable popular publications include *Radioactivity* (1904), *Radioactive Transformations* (1906), *Radioactive Substances and their Radiations* (1913), *The Electrical Structure of Matter* (1926), *The Artificial Transmutation of the Elements* (1933) and *The Newer Alchemy* (1937).

In 1900, Rutherford married Mary Georgina Newton (1876–1954) to whom he had become engaged before leaving New Zealand. They married at St. Paul's Anglican Church, Papanui in Christchurch and they had one daughter, Eileen Mary (1901–1930), who married, the astronomer and physicist, Sir Ralph Howard Fowler FRS (1889 – 1944). In 1937, Rutherford suddenly died from an untreated, strangulated, hernia died. He was cremated and his ashes laid to rest in Westminster Abbey. He is also commemorated by the naming of the, artificial, radioactive element, $_{104}$Rf, rutherfordium.

Rutherford had been due to speak at the opening session of the 1938 Indian Science Congress, and was replaced by Professor Sir James Jeans FRS who, in tribute, said *'In his flair for the right line of approach to a problem, as well as in the simple directness of his methods of attack, he often reminds us of Faraday, but he had two great advantages which Faraday did not possess, first, exuberant bodily health and energy, and second, the opportunity and capacity to direct a band of enthusiastic co-workers. Great though Faraday's output of work was, it seems to me that to match Rutherford's work in quantity as well as in quality, we must go back to Newton. In some respects he was more fortunate than Newton. Rutherford was ever the happy warrior - happy in his work, happy in its outcome, and happy in its human contacts.'*

Johannes Rydberg

Halmstad – born **Johannes *'Janne'* Robert Rydberg** (1854 – 1919) is renowned for devising the eponymously named formula by which he rationalised the Balmer and other series of the hydrogen emission spectrum.

$$\frac{1}{\lambda} = R_H \left(\frac{1}{n_1^2} - \frac{1}{n_2^2} \right)$$

where λ is the wavelength (in metres), $R_H = 1.097376 \times 10^{-7}$ m^{-1}, n_1 and n_2 are integers which can never be the same value. Following Bohr's theory of the atom it became apparent that n_1 is the higher energy level and n_2 is the lower energy level.

Rydberg studied at Lund University where he obtained his bachelor's degree and was awarded his doctorate. In 1879, Rydberg was appointed an assistant lecturer before being appointed an associate professor in mathematics in 1880, and in physics in 1882.

It was at this time that he began studying the standardised atomic weights of the elements due to his curiosity about the seemingly random increase in weights moving down and across Mendeleev's Periodic Table.

Rydberg struggled for several years to create a formula which would predict atomic weights and failed completely.

We now know that this is because, as Moseley demonstrated the atomic weight, as the relative atomic masses is not the fundamental property of the atom.

After several years, Rydberg turned his attention to the, newly discovered, atomic emission spectra.

This was much more successful and arose from the work of Johann Jakob Balmer who presented an empirical formula for the visible spectral lines of the hydrogen atom in 1885. Rydberg's research enabled him to publish a formula in 1888 which could be used to describe the spectral lines not only for hydrogen but also for other elements. The formula is displayed above and there are only two conditions:

- n_1 and n_2 are small integers
- and $n_1 \neq n_2$

Unable to obtain a fulltime academic position, Rydberg worked as an actuary until he was finally appointed to be a full professor in 1909.

In 1913, Rydberg became very ill and retired on the grounds of ill health in 1915 and he died in 1919.

Erwin Schrödinger

Vienna – born **Erwin Rudolf Josef Alexander Schrödinger** (1887 – 1961) is most renowned for his wave equation which formed the foundation for the entire discipline of quantum mechanics in which he described the wave function of the hydrogen atom.

This led to the concept of atomic orbitals and which reconciled the Bohr theory of the atom with the Rutherford concept of the atomic nucleus, Heisenberg's *Uncertainty Principle* (which states that it is impossible to know precisely the location and momentum of an electron and de Broglie's concept of wave – particle duality. For these insights he shared the 1933 Nobel prize in Physics with Paul Dirac.

He was, however, interested in far more than quantum mechanics and researched and wrote on a statistical mechanically interpretation of thermodynamics, electrodynamics, the physics of dielectric materials, the theory of colour, relativity and cosmology.

He is also renowned for his, 1935, **Schrödinger Cat** thought experiment which attempted to address the confusion arising from the theoretical concept that if electrons could be described as waves then the waves could superimpose on each other and this superposition would only collapse upon observation. This arose from the 1930 Copenhagen Conference which attempted to create an acceptable interpretation of all the confusing theoretical and experimental discoveries of the previous twenty years. Essentially, Schrödinger created an analogy where a cat was in a box which also contained a vial of poison and radioactive source. If an internal meter e.g. a Geiger counter detects the radioactive emission from the decay of a single atom then the vial is shattered, releasing the poison, which kills the cat. The Copenhagen Conference's interpretation of quantum mechanics implies that the cat is simultaneously alive and dead and only remains alive or dies when the box is opened.

This is an attempt to explain how electrons can exist in different waveforms until observed which is absolutely mind-boggling since the question then arises as to how an electron realises it is being observed. There is, to this day, no satisfactory explanation for this conundrum and neither is there any explanation for *quantum entanglement*, termed by Einstein as '*spooky action at a distance*'. *Quantum entanglement* arises when a pair of particles such as a pair of electrons, one spinning up and one down are separated. If the spin of one electron is reversed then the other particle also instantaneously reverses its own spin. This implies that the particles communicate at speeds faster than that of light which is theoretically impossible and also begs the question '*How does one particle know what its, separated, paired particle is doing?*'

The only child of an Austrian botanist, Rudolph, and a half – British mother, Georgine, who was the daughter of a renowned chemistry professor, Alexander Bauer, Schrödinger studied in Vienna and became an experimental assistant in 1910.

During the First World, he was a conscripted artillery officer in the German Army but after the Armistice he became an associate professor in Stuttgart before being appointed to a full chair at Breslau (now Wroclaw) before, in 1921, moving to the University of Zürich.

In 1927, Schrödinger succeeded Max Planck at the Friedrich Wilhelm University in Berlin and remained there for six years.

Vehemently anti-Nazi, Schrödinger left Germany in 1933 and became a Fellow of Magdalen College, Oxford shortly before he received his Nobel Prize, shared with Paul Dirac..

He was not made welcome at Oxford because of his unusual personal arrangements where he lived in a ménage à trois with his wife, Annemarie (Anny) Bertel, and his mistress, Mrs. Hilde March (the wife of a colleague), both of whom had left Germany with him. Consequently, he moved for short periods to chairs in Physics at both University of Graz and Allahabad University.

After the annexation of Austria by Germany, the Anschluss, Schrödinger suffered harassment because of his well known opposition to Nazism and, dismissed from Graz but ordered to remain in the city, he and his wife and his mistress fled to Italy.

During the Nazi years, the Republic of Ireland remained steadfastly politically and publicly neutral but Schrödinger received a personal invitation from Éamon de Valera, the then Taoiseach and a mathematician himself, to move to the country and establish the Institute for Advanced Studies in Dublin. de Valera personally authorised the visas for Schrödinger's wife and mistress to accompany him and, in 1940, Schrödinger accepted the invitation and became Director of the School for Theoretical Physics, remaining there for seventeen years. In 1948, he became a naturalized Irish citizen whilst still retaining his Austrian citizenship and in those seventeen years he published at least fifty more papers including concepts of the unified field theory, attempts create a uniform theory of the strong and weak electromagnetic forces, electromagnetism and gravity. During his time in Dublin, Schrödinger fathered another two children by another two women.

In his, 1944, book 'What Is Life?' Schrödinger also addressed genetics, looking at the existence of sentient life from the point of view of physics and discussed philosophical aspects of science, ancient philosophical concepts, ethics, and religion. Brought up Christian (Lutheran), he became an atheist but was fascinated by Eastern philosophies and religions and used much such symbolism in his writings. Two of the three recipients of the 1962 Nobel Prize in Physiology or Medicine, Francis Crick and James Watson both wrote independently that they had been inspired by Schrödinger's book to investigate the structure of DNA.

In 1955, Schrödinger retired from his position in Dublin and returned to Vienna to become an emeritus professor of physics at the city's university. He died of tuberculosis in 1961 and was buried in Alpbach.

At the time of writing, 2020, Schrödinger's grandson Professor Terry Rudolph is a professor quantum physics at Imperial College, London.

Sir George Paget *G.P.* Thomson

Sir George Paget Thomson, FRS (1892 – 1975) was the recipient of the 1937 Nobel Prize in Physics for his discovery of the wave properties of the electron.

He was the son of Sir J.J. Thomson, the discoverer of the electron and who was the recipient of the 1906 Nobel Prize in Physics and Rose Elizabeth Paget. Through the maternal line he was also grandson of the physician Sir George Edward Paget, FRS (1809 – 1892), hence his middle name. One of the patients of his grandfather was James Clerk Maxwell.

Thomson went up to Trinity College, Cambridge to study mathematics and physics but at the outbreak of the First World War he was commissioned into the Queen's Royal West Surrey Regiment with whom he served briefly in France before transferring to the Royal Flying Corps whereafter he researched aerodynamics at the Royal Aircraft Establishment at Farnborough.

Thomson resigned his commission, in 1920, leaving as a captain, and returned to Cambridge where he was elected a fellow before moving to the University of Aberdeen where he discovered the wave properties of the electron. Interestingly, whereas his father described electrons as particles, Thomson discovered the wave properties. Between them father and son were awarded Nobel Prizes in Physics for work on the same material.

In 1930 Thomson was appointed to a chair in physics at Imperial College London where he specialised in nuclear physics and in the war years, the military applications of the discipline. He remained at Imperial until 1952 when he became Master of Corpus Christi College, Cambridge.

It was in Aberdeen that Thomson met and married Kathleen, daughter of The Very Reverend Sir George Adam Smith, the then Principal and Vice-Chancellor of the University. Tragically Kathleen died in 1941 leaving Thomson with four children, John, David, Clare and two-year-old Rose, to raise. David married Patience Bragg, daughter of Sir William Lawrence Bragg, one of his father's closest friends and his grandfather's successor, with Rutherford in between, in the Cavendish Chair of Experimental Physics. Thus two families of huge achievement became united and so the children of David and Patience had two Nobel prize winning grandfathers and two Nobel prize winning great grandfathers, a unique circumstance. Another of the children of Sir George Adam Smith was Janet Adam Smith who wrote the standard biography of the great Scottish writer, politician and soldier, John Buchan.

G. P. Thomson was a very modest man and wrote comparing his work with that of Davisson and Germer:

'If one compares the two sets of experiments it is clear that those of Davisson and Germer were great triumphs of experimentation, among the greatest ever made. Those at Aberdeen were singularly simple and easy, the only serious difficulty being to prepare good specimens. It is not hard to see the reason for this difference. Davisson had discovered an effect which he thought might be important in a particular way. He had to study this effect more or less as he had found it. If he did a wholly different experiment it would neither prove nor disprove this explanation of what he had found.

I was trying to confirm the truth (or otherwise) of an idea based on a very general theory and was entitled to try the easiest experiment which could reasonably be expected to do so.'

Sir Joseph John *J. J.* Thomson

Manchester – born Sir **Joseph John Thomson** OM PRS (1856 – 1940) discovered the electron, in 1897, from his investigation into cathode rays for which he was awarded the 1906 Nobel Prize in Physics. He was the father of Sir George Paget Thomson FRS who determined the wave nature of the electron.

From his experiments he calculated that the electron must be much smaller than atoms and have a very high charge/mass ratio.

He is also credited with finding the first evidence (1913) for isotopes when he was investigating stable elements to study the canal rays (positive ions) which had first been detected by Eugen Goldstein in 1886. Thomson supervised the work of F.W. Aston who devised the first mass spectrometer, which he termed a mass spectrograph as the original detector was a photographic plate.

Born into a solid middle class family, Thomson was educated privately and having shown prodigious mathematical talent was admitted, aged fourteen, to Owens College which is now part of the University of Manchester. Thomson moved to Trinity College, Cambridge, in 1876 to study mathematics and was awarded his Bachelor of Arts (B.A.) degree in mathematics in 1880. He was elected to a fellowship at Trinity in 1881 and was awarded his M.A. in 1883.

It is amusing to learn that, although renowned as a experimental physicist, Thomson was notoriously clumsy and was deterred from performing the experiments himself since he allegedly always broke the apparatus. This resonates with the '*Pauli effect*' **which was, affectionately, named after it was noticed that, very often, experimental apparatus broke, merely apparently because of Pauli's presence.**

In 1884, Thomson was appointed Cavendish Professor of Physics at Cambridge and remained at Trinity College and at the Cavendish Laboratory for the entirety of his career.

Renowned as a great teacher and supervisor, one of Thomson's students was Ernest Rutherford who discovered the nucleus and also supervised the work of six students who later became Nobel Laureates in Physics (Charles Glover Barkla, Niels Bohr, Max Born, William Henry Bragg, Owen Willans Richardson and Charles Thomson Rees Wilson) and two who were awarded the Nobel prize in Chemistry (Rutherford and also F.W. Aston). A Nobel laureate himself, Thomson's son, George Paget Thomson, was also a Nobel prize winner in Physics (1937).

In his early research, Thomson studied the electromagnetic wave theory proposed by James Clerk Maxwell, introduced the concept of the electromagnetic mass of a charged particle and also demonstrated that a moving, charged body appeared to increase in mass whilst in motion. His book, '*Notes on recent researches in electricity and magnetism*' (1893), built on Maxwell's theories so well that it was often described as '*the third volume of Maxwell*', Maxwell having, of course, written the first two volumes himself. Thomson's third book, '*Elements of the mathematical theory of electricity and magnetism*' first published in 1895 was a popular and readable introduction to the subjects and became a standard textbook.

Thomson was the not the first to hypothesise that atoms contained more fundamental particles. For example, William Prout and Norman Lockyer had proposed that the smallest unit was the hydrogen atom and that all larger elements comprised multiple numbers of hydrogen atoms. Until the discovery by Aston of isotopes, this remained inexplicable not least because the atomic weights of the elements (as the relative atomic masses were then termed) were rarely of integral values.

Thomson made his initial suggestion of the tiny size of electrons, which he termed cathode rays, following his discovery that the rays travel much further through air than would be expected for an atom-sized particle. Thomson estimated the mass of cathode rays by measuring the heat generated when the rays hit a thermal junction and he compared this with the magnetic deflection of the rays. He concluded that cathode rays were at least 1,000 times lighter than the hydrogen atom and also that the mass of these particles remained the same in whichever atom they existed.

He concluded that the rays were composed of very light, negatively charged particles and were a fundamental building block of the atom. Thomson called the particles 'corpuscles' which was the term Isaac Newton had used when he believed that light was comprised of particles. Over time, however, the term electron became the accepted name for the tiny particle.

One reason for Thomson's conclusion that the electrons were fundamental i.e. indivisible particles was that, in contrast to the anode rays (canal rays) they were always deflected as a single beam whereas the anode rays produced a spread of signals. Since all cathode rays behaved identically in all circumstances they must all be the same.

Moreover, since the anode rays produced a multitude of signals, he also reasoned that the anode rays were multiples of the one of lowest mass/charge ratio and this led him to his 'plum pudding' model of the atom.

In 1905, Thomson reported the discovery the natural radioactivity of potassium and, in 1906, he demonstrated that hydrogen possessed only a single electron per atom.

In 1890, Thomson married Rose Elisabeth Paget who he had met when she had observed his public demonstrations and attended his public lectures. She was the daughter of Sir George Edward Paget, a physician and who was Regius Professor of Physic at Cambridge. This position is one of the oldest professorships in any university anywhere, having been established by King Henry VIII in 1540 and the appointment has to be approved by the monarch of the time. At the time Cambridge did not permit female undergraduates, postgraduate or female lecturers and, indeed, female students were not awarded degrees until 1948.

Thompson and Rose Paget had two children, the aforementioned Sir George Paget Thomson FRS and Joan Paget Thomson (later Charnock) who was a prolific author, writing childrens' books, non-fiction and biographies.

Thomas Young

Somerset – born **Thomas Young** FRS (1773 – 1829) was a polymath who became renowned for his contributions to the study of light and vision, mechanics, energy, physiology, musical harmony and to Egyptology, contributing to the deciphering of the Rosetta Stone.

Young's physics experiments both informed and formed the basis of later work by William Herschel, Hermann von Helmholtz, James Clerk Maxwell, and Albert Einstein.

Young is most famous for establishing the wave nature of light which apparently disproved the particle *'corpuscular'* theory of Sir Isaac Newton and his discoveries were later verified by Augustin-Jean Fresnel.

Astonishingly talented, Young, the scion of a Quaker family in Milverton, had, by the age of fourteen, learned Latin and Greek and had developed a more than passing acquaintance with the French, German, Italian, Hebrew, Aramaic, Syriac, Samaritan, Arabic, Persian and Turkish languages.

In 1792, he went to St. Bartholomew's Hospital to study medicine before moving to the School of Medicine at the University of Edinburgh in 1794, progressing a year later to the University of Göttingen where he was awarded his degree of doctor of medicine in 1796.

He then proceeded to Emmanuel College, Cambridge in 1797 and became financially independent following a large inheritance from his great uncle. By 1799, he had established himself as a physician at 48 Welbeck Street, London W1.

Between 1801 and 1803 Young was a professor of natural philosophy at the Royal Institution and in those two years he delivered ninety one public lectures which were published, in 1807, as *'A Course of Lectures on Natural Philosophy'*.

For our purposes, Young's greatest achievement was to determine the diffraction of light which contradicted Newton's corpuscular, particle, theory of light and demonstrated its wave properties.

In an 1804 paper, *'Experiments and Calculations Relative to Physical Optics'*, Young described an experiment in which he placed a card measuring approximately 0.85 millimetres in a beam of light from a single opening in a window and observed the fringes of colour both in the shadow and to the sides of the card.

He reported another observation that placing another card in front or behind the strip which prevented the light rays from striking one of its edges caused the fringes to disappear which, in analogy to ripples of water, supported Young's contention that visible light is composed of waves.

He reported that the same observation applied to reflections from thin films of soap and of oil.

Within ten years, Young's observations were reproduced, reported and extended by Augustin-Jean Fresnel.

Glossary

The following brief glossary clarifies terms which are used elsewhere in this volume.

It is not intended to be anything more than a very concise summary of concepts relevant to this work.

Glossary

Absorbance

Absorbance is a measure of the capacity of a substance to absorb light of a particular and specified wavelength and is calculated according to the formula:-

$$\textbf{Absorbance} = 2 - \log_{10}(I / I_o)$$

where I is the intensity of the light passing through a substance and I_o is the intensity of the incident light.

Alkali

A definition of **alkali** depends on the definition of a *base* which is a substance that reacts with, and neutralises, an acid and are usually metal oxides, hydroxides, carbonates or hydrogencarbonates. An **alkali** is any *base* which dissolves in water.

Alkane

An **alkane** is any hydrocarbon of the formula C_nH_{2n+2} where n is any integer

Alkene

An **alkene** is any hydrocarbon of the generic formula C_nH_{2n}

Alkyl group

An **alkyl group** is any alkane attached to a carbon atom and has the generic formula: C_nH_{2n+1}

Alkynes

An **alkyne** has the generic formula C_nH_{2n-2}

Alkynes are extremely reactive and the only one of any real interest or use is ethyne, commonly known as acetylene: $H - C \equiv C - H$

Alpha particles

Alpha particles were first detected through the discovery of natural radioactive and contain two protons and two neutrons. This means that they have a 2+ charge and are frequently described as helium nuclei. This often confuses people since no atoms of helium may have been present. They are better described as identical to helium nuclei.

Anode

The **anode** is a positively charged electrode and will attract negatively charged ions towards itself. Consequently the negative ions are called anions.

Atom

The **atom** is the smallest part of an element which behaves like the bulk element.

Beta, β, particles

β particles are electrons

Atomic mass unit

In the 19th century, elements were ordered by, the now archaic term, atomic weight. It was decided that since hydrogen was the lightest element then all the weights of all other elements were recorded relative to hydrogen which was assigned an **atomic mass unit** (AMU) of one. In 1961, it was decided that the term atomic weight was inconsistent with physics and should be changed such that one atomic mass unit (AMU) was defined as 1/12 of the mass of a 100% pure sample of the ^{12}C isotope.

Charge

Charge is the ability of a material or object to attract or repel another substance or object. The concept of charge has been known for thousands of years but it has only been in the last one hundred years that it has been understood to be caused by the gain or loss of electrons.

Diffraction

The process by which a ray of waves is spread out or *dispersed* as a result of passing through a narrow aperture or across an edge. This is termed **diffraction** and the easiest, visual, example is of water ripples where water, forced through two holesor slits, creates regions which are completely calm (flat) and regions where the height of the waves is greater than the height of the initial wave. This also applies to light and Thomas Young proved the wave-like nature of light by producing a result similar to those of water waves with alternative regions brighter than that of the initial source and also completely dark regions. Shooting a series of balls i.e. particles would not achieve the same effect.

Element

An **element** is a substance which cannot be broken down into anything simpler by chemical means. This qualification is important since, of course, an element can be smashed apart using physical means.

Empirical Formula

The **empirical formula** of a compound is the simplest and lowest ratio of the elements whilst the molecular formula records the exact number of each type of atom present. The clearest example is that of ethane which has the molecular formula, C_2H_6, indicating that each molecule comprises two carbon atoms and six hydrogen atoms. Clearly, this ratio is 2:6 but a simpler ratio is half of that so the empirical formula of ethane is CH_3.

Energy

Energy is closely related to work. In physics terms, work is conducted and completed when something is made to move. Energy is defined as the ability to perform the defined work and there are two types of energy.

Kinetic energy (KE) is the energy associated with movement.
Potential energy is stored energy and can comprise chemical energy, gravitational energy as well as nuclear, electrical, magnetic energy etc;

Fluorescence

Fluorescence is the emission of light by a material which has absorbed higher energy electromagnetic radiation or another form of energy. The emitted light is of longer wavelength and hence lower energy than the absorbed radiation. The phenomenon has long been known but it was not understood and is most striking when the material absorbs the, invisible to the human eye, ultraviolet radiation and responds by giving out visible light of particular colours. The phenomenon is beyond the scope of this series but it becomes relevant when we consider that the discovery of the nucleus was only possible because alpha particlesimpinging on a zinc sulfide screen, the detector, caused flashes which were counted the locations recorded.

Frequency	**Frequency** is the measure of any number of events in any particular time period. Conventionally, in physics, frequency is the number of events per second and, with regard to electromagnetic radiation is the number of waves passing a particular location every second. The units of frequency, cycles per second, are denoted as Hertz (Hz) and the numbers can be huge and so become abbreviated as Megahertz (MHz) – the number of millions of events per second and even Gigahertz (GHz) – the number of billion of events per second.
Functional Group	A **functional group** can be regarded as a group of atoms which, essentially, behaves a one atom and which behaves the same regardless of the molecule to which it is attached. Examples are alkyl, hydroxyl (– OH) and carbonyl (C=O) groups.
Ground state	The **ground state** is the lowest possible energy level and is the natural state in which electrons exist within an atom. When an electron is promoted to a higher energy level the atom is said to be in an *excited state* and the promotion of the electron is described as excitation. The simplest analogy is the first step of a ladder.
Hybridisation	The concept of orbitals led to huge advances in the understanding of covalent bonding but the electronic configurations of the carbon atom and the hydrogen atoms did not explain why methane (CH_4) possesses four C – H bonds of equal length and, in other words, equal strength. Linus Pauling explained the process by proposing that, in the case of carbon, the overlapping 2s and 2p orbitals mix themselves up to produce four equivalent orbitals, known as sp^3 orbitals. The sp^3 – orbitals resemble a squeezed party ballon as shown right:

The final number of hybrid orbitals must be the same as the original number and so the mixing of one s – orbital and three p – orbitals must create four orbitals of identical size and shape and this accounts for the shape of the methane molecule and the identical lengths of the four bonds as shown below.

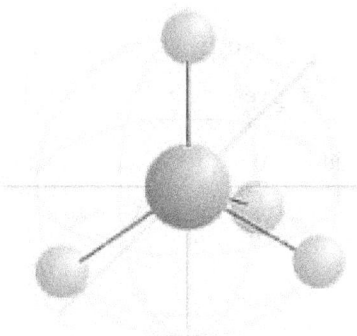

Integration	Mathematically **integration** is the area under a curve and the resulting value is known as the integral.
Intensity	**Intensity** is the quantitatively stated brightness of a ray of electromagnetic radiation.
Ion	Chemically, an **ion** is an atom which has either gained or lost one or more electron and so is charged. Ions were, however, known before the existence of the electron was known since it was determined that some materials were attracted to the positively charged pole of an electric field and others to the negative pole. Since the plates were charged then there had to be a movement of ions not least because in many electrochemical experiments one electrode gains mass whilst the other loses mass demonstrating beyond doubt that there was a movement of material.
Ionisation	**Ionisation** is the formation of an ion by the loss or gain of one or more electrons from the atom. The loss can be caused by irradiating materials with ultraviolet light or radiation of greater energy and that loss is known as ionisation.
Isoelectronic	**Isoelectronic** atoms or ions are those which, irrespective of the atomic number, all contain the same number of electrons. For example the lithium atom contains three electrons with two in the lowest energy level (n=1, where n is the numbering of the energy levels) and one in energy level n=2. As explained in the main text, lithium loses one electron to form the $_3Li^+$ ion which contains two electrons as does the helium atom. This means that the lithium ion and the helium atom, both containing two electrons, are isoelectronic.
Isomers	**Isomers** are molecules of the same molecular formula i.e. which contain the same number of each type of atom but behave differently.
Isotopes	**Isotopes** are atoms of the same atomic number but different numbers of neutrons such as those of hydrogen: ^1H, ^2H and ^3H all of which contain one proton and, respectively zero, one and two neutrons.
Molecule	A **molecule** is a group of atoms bonded together which represents the smallest fundamental unit of a chemical compound that can take part in a chemical reaction.
Molecular Formula	The **molecular formula** states which elements are present and the numbers of each type of atom present in the discrete molecule.
Momentum	**Momentum** is the product of an object's mass(m) and velocity(v). Momentum (p) = mass (m) x velocity (v)
Neutral	A **neutral** substance is one which is not attracted to a positively or negatively charged electrode and is completely unaffected by an electric field.

Photobiology   The study of the effects of of visible light and lower energy electromagnetic radiation on living organisms, both plants and animals.

Photochemistry   The study of a chemical reaction caused by absorption of ultraviolet light, visible light or infrared radiation.

Photon   A **photon** is a packet of electromagnetic energy.

Picometre   One **picometre** is equivalent to 10^{-12} m.

Qualitative   A **qualitative** description is a measure of some property or quality of an entity such as its colour e.g. silver and gold are readily distinguished.

It gives no indication of a physical property such as width, diameter, volume etc; where the exact dimension is quoted but it can be used to compare in non-quantitative ways such as larger/smaller, brighter/dimmer, stronger/weaker, black/white, blue/green etc;

Quantitative   A **quantitative** measure is one where the property of a substance or object is expressed numerically.   Examples include length, volume, mass etc;   In science, the quantitative property is expressed with its units of measurements but mathematically this is not necessary since numbers are, by definition, quantitative but not always possess units.

Quantum   A **quantum** is a, particulate, packet of energy as described by Max Planck and the concept was used by Albert Einstein to explain the photoelectric effect.

Resonance   In physics, **resonance** describes the increased amplitude of a wave or a system that occurs when the frequency of a periodically applied force is equal or close to a natural frequency of the system on which it acts or when a system responds to particular energies.

Valency   **Valency** was defined by Sir Edward Frankland as the combining power of an element and can be described as the number of hydrogen atoms it can displace or combine with. By this definition, then, hydrogen has a valency of one, oxygen as a valency of two, nitrogen a valency of three and carbon a valency of four as exemplified by the molecules of hydrogen (H_2), water (H_2O), ammonia (NH_3) and methane (CH_4) respectively.

Wave   A **wave** is a means of transferring energy from one location to another without movement of any matter.

Wavelength   The **wavelength** is the distance between identical locations on two adjacent waves.

Wavenumber   The **wavenumber** is the number of waves per centimetre with the simple units cm^{-1} since the number of waves has no units.

Index

Personalities

Locations